THE CONFETTI GENERATION

The CONFETTI GENERATION

How the New Communications Technology Is Fragmenting America

WILLIAM J. DONNELLY

HENRY HOLT AND COMPANY NEW YORK

Published by Henry Holt and Company, Inc.
521 Fifth Avenue, New York, New York 10175.
Published simultaneously in Canada.

Library of Congress Cataloging in Publication Data
Donnelly, William J.
    The confetti generation.

    Bibliography: p.
    Includes index.
    1. Mass media—Technological innovations.
2. Mass media—Social aspects—United States.
3. United States—Popular culture. I. Title.
P96.T42D6   1986   302.2'34'0973   86-7552
ISBN 0-8050-0095-X

First Edition

Designed by Jeffrey L. Ward
Printed in the United States of America
10 9 8 7 6 5 4 3 2 1

Permission granted from *Newsweek* to quote from "The Druse,
Alawites and New Jewels" by Meg Greenfield, *Newsweek,*
December 26, 1983.

ISBN 0-8050-0095-X

To Patricia

Contents

Preface

During the 1970s, when I was at Young & Rubicam, the world's largest advertising agency, part of my work involved projecting the future of communications and the impact new media might have on the business of the agency's advertising clients. In those years we had a rather straightforward perception of this task. Clients used print and broadcast media as advertising vehicles. If new media were coming along, they would change the usage patterns of existing media and would offer new advertising opportunities. This was important from a planning point of view because of the billions of dollars spent on advertising and the role of advertising in generating the gross national product of American business.

In 1976 Y & R published the work I had done on these issues under the title *The Emerging Video Environment*. My projections were quoted widely in books ranging from Alvin Toffler's popular *The Third Wave* to college textbooks such as *Understanding Mass Communications* by DeFleur and Dennis, and the projections proved remarkably accurate. Y & R clients put their money where my mouth was and became leaders in supporting cable television and the other new media.

I had always perceived myself as a maker and marketer of communications product, as well as a thinker about communications. A general American prejudice suggests that you can be a marketer or a thinker, but you cannot be both. This may be generally true, but when I joined Y & R in 1972, I bet that it was the only place where one could deal with all the media at once on both practical and conceptual levels. I believed then and still do that to predict and manage the future in an athletic and practical way requires a realistic grounding in the creative, business, and conceptual disciplines.

From the first days of television, I was intimately involved in diverse phases of communications, from the practical to the theoretical, as witness and active participant. I witnessed my father's work manufacturing television tubes, and I myself worked as a printer. While an instructor in communications at Loyola University, Los Angeles, I put KXLU, an FM radio station, on the air. I produced and directed theatrical plays and performed on radio and television. In the early days of Public Broadcasting, I founded and edited a trade magazine called *Educational/Instructional Broadcasting.* I published *Gabriel,* a quarterly magazine for the Catholic Broadcasters Association, and for eight years wrote a column of television and film criticism that was eventually syndicated in forty-one newspapers. I have written film and television scripts. I was the news director for KCET-TV, the Public Broadcasting station in Los Angeles. When I returned to New York in 1970, I founded a newsletter now called *The Home Video Report,* which reported on the first labor pains of the new electronic media. After I joined Y & R, we published a book called *Video Cassettes: Medium, Market, Systems, Programming,* which in its day functioned as a key sourcebook for the industry.

Thus, at the dawn of the video revolution and when I joined Y & R, I already had a substantial grounding in all aspects of media both old and new. This gave those of us at Y & R the confidence to project what would happen next. This was the background from which for a decade we conceptualized, funded, and created in all areas of the new media, from cable to cassettes to videotex.

But there was something missing in all our prophecies—the people—the fathers, mothers, and children who would use the new technologies. When asked to project the next decade in 1980, I realized that people create markets and that the critical path for advertisers would be changes in our needs and wants brought about by new technology. This led to my first projection of the Confetti Generation, published as a pamphlet by Y & R in 1982.

Worldwide speaking engagements offered me the opportunity to share these concepts with a wide variety of people, and in the process I learned something. I learned that on a scholarly level no one had quite identified how communications technologies relate to our behavior and how they can predict our social character. I learned that no one had yet collected the facts about the new media in one place. What facts existed were not readily accessible, and eclectic speculation, it seemed to me, was detrimental. I also learned that people cared more about their own lives than about their businesses; and, as they accepted the projection of the Confetti Generation, they wanted to know how they might prepare themselves for personal survival during it. This was a lot to think about.

Thus in 1983 I resigned as a senior vice-president of Young & Rubicam to write this book. I knew there were many ways to fail, but I also knew that I couldn't think about and construct some answers to the remaining questions while executing the businesses I managed. My wife, Pat, supported my monastic choice, but only other people will be able to judge if I "got it right" and whether the effort was worthwhile.

THE
CONFETTI
GENERATION

Introduction

We live "fast forward." As a people we are generally optimistic about the future and rush toward it. That very American general "Vinegar Joe" Stilwell once said, "We can't expect to be told about the future. If we want to find out, we must march toward it." And that is what Americans have always done. We experience occasional brief flurries of concern that perhaps we are changing the world we live in faster than we can properly or happily change ourselves. But, on the whole, we are a confident, Promethean society believing that change based on human invention is both inevitable and better.

The Electronic Revolution

We are poised today on the edge of radical developments in communications technology. The new electronic media—which include communications satellites; broadcast, cable, and pay television; videocassettes; video games; videodiscs; interactive videotex services; and industrial, personal, and home computers—have not been kept a secret from anyone. Buffed to the patina of science fiction, they provoke interest, tantalize inquisitive minds, agitate both protectionist lobbies and would-be regulators, stimulate entrepreneurial imaginations, and build consumer expectations. Almost every magazine we read carries articles on the profusion of products and services promised by the new electronics. This is today's frontier, the one place—besides space exploration—where we do not seem to be faced by limits.

Born in an age of tinkerers, America grew up in the centuries of scientific expositions and world fairs. It has become a national pastime to gawk at, admire, tinker with, and dream about new technologies. Our fascination with the new electronic media is simply the latest chapter in the history of American ingenuity and know-how. We recognize and even embrace what seems to be our destiny—namely, that we will change the world we live in. But we seldom comprehend how these technologies will change our lives on the most personal and intimate levels, how they will change the way we think, believe, feel, and behave.

Of course, our understanding of technology is much more sophisticated today than it was in the past. We know that new technologies change societies and the people who live in them. We even have an army of people paid to think about the future at institutes and universities, as well as a proliferation of journalists constantly reporting on the latest inventions and the consumer products they will spawn. Yet the speculation of intellectuals and reportage of journalists offer very little information that we can put to use in our personal lives. A huge gap exists between the formulations of the high-tech mandarins and the glossy "quick reads" of headline-conscious journalists. It is to bridge this gap and to sharpen our personal imaginations of the future that this book has been written.

The consensus among the high-tech mandarins, both institutional and popular, is that the new communications herald a new wave of social and economic change and that they will create an information society fueled by information industries. There will be a new economic of knowledge, and we will live in a global village more pervasive than the one created by television and introduced to us by Marshall McLuhan. This new global village will be determined by powerful, ubiquitous computers and by satellites linking them together. Economically, this global village will be a reflection of expanding multinational corporations. Necessarily, our occupations will change, and society too will change in an upward spiral of economic evolution.

These prophecies are not all about money, of course, or about the global pursuit of wealth as an end in itself. Think-tanker Herman Kahn asserted in a paper published by the Hudson Institute in January 1982, "The coming boom in technology will be considerably more than just an economic phenomenon—it will also help foster a constructive synthesis of old and new values, a resumption of the U.S. role as the leading source of technological innovation, and a new sense of excitement about technology and science and of being proud to be American." This booming optimism—or chauvinism, if you will—in the humanizing impact of expanding economies parallels Japanese prime minister Yasuhiro Nakasone's belief, as expressed in his 1983 speech to the Japan Society, that an "information society" will "also enable people to satisfy their quest for personalized values and new life-styles within our modern industrial society."

There is little doubt that the economies of the industrialized world are becoming more information-oriented and that more and more of their income will be derived from the generation, processing, and analysis of data. Not only will new knowledge industries be developed, but information handling with the aid of computers will probably be the way in which we conduct much of our business and do much of our work. In the process, geopolitical borders will erode and previously separated business functions will merge. All of this is very heady and exciting, and very important to corporate planners and government policymakers.

When it comes to our personal lives, however, this heavy, intellectual, strategic planning accomplishes little but to make us fearful about our jobs, concerned about the type and quality of the education our children are receiving, anxious that our country is doing what it must to remain competitive and successful, and worried that all this data proliferation may erode our privacy and freedoms. Furthermore, high-powered speculation about the outside world and its institutions tells us little or nothing about our personal worlds, about our individual lives, about how we might think and choose in the future.

We are willing to believe in the descriptions of a high-technology world of the future, but in believing we also want to know what such a world really means in the context of our own lives. When we think about the future, we each think about our own personal future. The mandarins explore only the larger picture and lose sight of our individual concerns—the concrete details that are the stuff of our lives.

At the other end of the spectrum are best-selling journalists and gurus who are more than willing to tell us about these new technologies in our daily lives. They cavalierly disregard social thinker Daniel Bell's advice in *The Coming of Post-Industrial Society* to avoid forecasting things to come "which would have to be . . . an empirical set of observations." Alvin Toffler, for example, warns us that the future will shock us, but then proceeds to assuage our fears by telling us what we want to hear about shopping centers, schools, and household utilities. And John Naisbitt asserts quite specifically in *Megatrends* that "by the end of the 1980s cable television will offer as many as 200 different channels of entertainment and services," and that "cable television will be like the special-interest magazines: You will be able to tune in *Runner's World* or *Beehive Management*." Encouraging observations like these by various gurus and journalists make the future palpable and feed our optimistic consumer imaginations.

Yet something crucial is missing from these apparently empirical observations: They are not based on experience or structured observation. The defense, of course, is that these events and circumstances will take place in the future and, therefore, cannot be empirical in the strict sense. Yet we see far more disciplined and empirical efforts on the part of men and women involved in financing, manufacturing, and marketing the products and services that will create our communications future. These are the people who must determine in advance what to write, what to produce, what to manufacture, or what services to offer. For example, a marketing manager trying to identify the consumer benefits of a new technology, how it should be competitively positioned, how it should be priced, and what return on investment it might represent has a much more concrete sense of a communication technology's potential or probable future than someone who airily speculates that cable television will be offering programs like *Beehive Management* on two-hundred-channel systems by the end of the 1980s, or ever.

It is possible that in awarding a cable television franchise, some municipality somewhere, spurred on by its own fantastic imaginings, will demand construction of a two-hundred-channel system. But no one schooled in cable economics, no one responsible for the financing, construction, and profitable management of cable systems believes that two-hundred-channel systems will be prototypical in 1990, or that they could be anything but aberrant examples. Most media prophets display not the slightest evidence of having struggled with the programs and profits, systems and financing, and their predictions about the new electronics take on an ephemeral radiation, the square root of fantasy.

The problem of most so-called concrete speculation about the future of consumer communications is that it is presented as a series of headlines, alternately heralding success and failure as new technologies and businesses are born and encounter reality. The effect is a dialectic of "everything is possible/nothing is probable" that robs our imaginations of any factual grounding.

The result is fantasy. Since consumer communications will touch our individual lives and shape our culture as broadcast television did, we should not be so cavalier about them.

An understanding of how the new electronic media will affect our futures depends on a fact-based and involved understanding of the new electronics. Since most of the observations by journalists such as Alvin Toffler are ungrounded, random, and eclectic, their forecasts of things to come cannot provide the answers we need. As we struggle to see ourselves in front of these new technologies, trying them on for size, judging how they fit our lives and psyches, we need to feel the goods. If we believe everything, we cannot believe anything. Yet, without some particular beliefs in the future, we cannot shape it or influence it. Without some understanding, we may not even be able to march toward the future—it will march over us instead.

Lost Imaginative Space

We are thus faced with a gap. We all want to know what conditions we and our children will be living under. We want to know how and why we will choose and act in the future. We know that technology alters human affairs, and we are caught up in seizing the new technologies and positing a new frontier. But when we try to develop a concrete imagination of the future and the difference it will make to our inner lives, we are confronted with a conundrum.

At one end of the spectrum are the high-tech mandarins skirting the limits of the sociological imagination—while ignoring the fact that sociology and economics never encompassed it all, especially the most personal depths of human life. At the other end of the spectrum are the journalists and prophets of popular science whose work is so obviously the result of secondary research; like Sibyl in her cave, their endless index cards echo disconnected facts and pronouncements, some of which must come true. We end up confused and our imaginations are disoriented.

We desperately need a place to ground our imaginations. In the current dialogue we are torn between fears and hopes. We fear for our privacy—and for control of our lives. We fear ending up as statistics, in a bureaucracy where technocrats exploit their knowledge of our personal information. We fear that our jobs may be replaced by silicone chips, while we are left untrained for other work. We fear that the costs of new technology, data, and services, coupled with a new illiteracy, will widen the gap between the haves and have-nots. Our democracy may be threatened, and the nation may be headed for oligarchy or demagoguery or both. We fear that our families will fragment and the generation gap grow even wider. We fear communications that may erode our morality. We fear the complexity and power of the new communications.

But we also have hopes. We hope that our economy will be brought to life with new energy. We hope that new conveniences will produce more time to spend on ourselves and with others. We hope that we will be able to spread the wealth of education and adapt training to individual needs. We hope that new choices will create better entertainment for ourselves and our families. We hope that with improved communications, government will be more responsive and people more involved. We hope that we will be able to overcome separations and renew our nuclear families. We hope for greater security of health and home.

Most of the fears and hopes emerging from the current dialogue represent issues, problems, and options that are beyond our reach and apparently beyond our personal resolution. When we think about the world beyond us, we wait upon others outside ourselves for direction. When we consider our personal lives, we can only pray that we will make the right choices for ourselves, our children, our fellow citizens. We know that unless we control communications technology and its products, they will control us—or someone will control us through them.

At this point our imaginations fail us. For all that we have read and heard, we cannot get a handle on the control the communications technology exerts, on how the process works, and on how we might find ourselves within it.

The critical missing link between communications technology and our personal lives is culture. Society arises and exists through communications, and the product of communications within society is culture. As participants in society, we are both cause and result of our culture.

Culture includes the historical knowledge, the understanding of the natural world, the traditions, values, points of view, beliefs, and opinions that we share as members of society. Daniel Bell defines culture as the "effort to provide a coherent set of answers to the existentialist situations that confront all human beings in the passage of their lives." He means birth, death, love, justice, indebtedness, and interpersonal relations, as well as God, democracy, and the nature of outer space. Our attitudes are formed in our culture. Culture is the product of society communicating, and these communications are shaped by the inherent capacities of the technologies we all use to communicate.

We may balk at the implication that our individual personalities are caught up in the group experience called culture; but, like the mountain, it is there. In W. H. Auden's words, "common culture shapes the separate lives." As individuals we can no more avoid definitive interaction with our culture than we can avoid the weather. It represents both possibility and limit. Thus we have a vital interest in this meeting point between communications technology and ourselves.

There is little doubt about the revolutionary potential of the new electronic media in our society and culture. What is not yet clear is how quickly the revolution will take place. It may, in fact, appear more evolutionary than revolutionary. But the cultural effects of a communications technology can be in evidence long before the media are clearly ubiquitous. This was true of television and will be even truer of the new electronic media.

The new electronic media, taken together, may in fact be more revolutionary than television. And, because of their direct appeal to critical segments of society, total distribution will not be required to effect a cultural revolution. Nevertheless, at some very early point, nearly everyone between the ages of ten and forty-nine today will have at least two new electronic media in their homes: (1) a random access information system (home computer, videotex, teletext, optical videodisc); and (2) a fickle access entertainment system (cable television, pay television, videocassette, videodisc, video game). A combination of any two will be sufficient to radically transform American society.

What is so revolutionary about the new electronic media? The new media encompass a quantum leap in the ability to store and retrieve information; a quantum leap in the availability of mimetic entertainment that is readily internalizable; a quantum leap in services that provide controlled, individualized, remote transactions; and a quantum leap in speed, to the point of warping our sense of time. These quantum leaps are the inevitable scenario of the future. Taken together, they will produce and define the Confetti Generation.

The question confronting us is, How can we absorb such an explosion of information and entertainment, such an implosion of speed and remoteness? What we need are personal and cultural tools for harnessing, shaping, directing, absorbing, and consuming such energies. It is at this point that a rooted imagination of the future is required, for through our own activity we can control the new electronics, rather than allow them to control us.

PART 1

THE
PEOPLE
CONNECTION

The dynamic relationship between communications technology, our culture, and our personal lives can be revealed and understood by examining as a case study the experience of television over the last forty years. Because television happened so rapidly we can isolate its "before" and "after"; we can track the difference it made in our personal lives and attitudes. In Part I, we will explore how the nation moved from the Organization Man to the Me Generation, and the role television played in that process. If we can isolate the precise effects television technology had on our social character and the manner in which that happened, we will be in a position to project the changes the new electronic media will probably bring. Our purpose will not only be to project the future, but to acquire that imaginative grasp of our personal futures that has thus far eluded us.

Over three decades ago, in the early 1950s, we overlooked the staggering characteristics of television technology; its speed and natural pervasiveness, its mimetic attractiveness, and its easy internalization. Yet, had the technology been better understood, had television been given different direction within the culture, had the social contract between supplier and consumer gone beyond facile satisfaction, television's impact would have been dramatically different. There could have been more to the message than the medium. But because we failed to imagine the impact of a technology that was far more than the addition of pictures to the innocence and modest power of radio, the technology followed its own path, and the medium did become the message. Without a practical imagination of the future, our destinies are liable to be determined by outside forces and technology once again.

1

The Television Audience

Even though television is one of the most common facts of our common lives, it has so escaped our grasp that it has lived a life of its own from its very beginning. The story of how and why that happened forms the necessary foundation for our understanding of what might happen when the new electronic media come to pervade our lives.

Television: A Case Study

The first 178,000 sets were manufactured in 1947; less than ten years later more receivers had been sold than there were households in the United States. The fulcrum year was 1955. As early as 1950, 29 percent of our homes had television, and by 1960 television penetration had reached 80 percent. By that time, however, American consumers had purchased one-third more television sets than there were places to live.

When David Sarnoff, chairman of RCA, introduced television at the New York World's Fair in 1939, he called it "radio's world of tomorrow," for it would add "radio sight to sound." He predicted that "as an entertainment adjunct, television will supplement sound broadcasting by bringing into the home the visual images of scenes and events which up to now have come there as mind-pictures conjured up by the human voice." Ten years later, when L. E. Parsons erected an antenna in Astoria, Oregon, to bring Seattle television over 123 miles of mountains, and thus initiated what would become cable television, he explained his endeavor in pretty much the same terms. He wished to satisfy his wife's desire "to have pictures with my radio."

But television was more than radio with pictures. Once let loose, the pictures took on a life of their own. Television inherited its philosophy, talent, means of support, programming content, and government control from radio, but the technology itself went far beyond radio. As early as 1935, German film critic Rudolph Arnheim warned in *Film as Art* that with the arrival of television, "the detour through the describing word becomes unnecessary," and that television's provision for instant experience of a visual sort would encourage the tendency toward a "cult of sensory stimulation"—a cult that forty years later would be called the Me Generation.

What we believe, affirm, and assert about the purpose and properties of communications media is important because our beliefs and assertions affect not only our response to communications, but what each medium can or cannot be or do in our society. The first Motion Picture Code, for example, conceptualized films as an art form, and this gave films their particular purpose, function, and license. The code said: "The art of motion pictures has the same objectives as the other arts—the presentation of human thought, emotion, and experience in terms of an appeal to the soul through the senses." Television, on the other hand, was perceived quite differently.

The Preamble to the Television Code first focuses on the home, then leads to the singular conclusion that "it is the responsibility of television to bear constantly in mind that the audience is primarily a home audience, and consequently that television's relationship to the viewers is that between guest and host." Rather than fulfill the traditional purposes of the arts, television was to be a congenial raconteur who came to dinner—quite a different role. The expectations of the audience—their conception of the medium—limit both the style and content of what can be communicated through the medium. Precisely because the suppliers and customers of television have an agreement on conception and intention, so much television criticism premised on artistic standards and intentions is irrelevant and falls on deaf ears. The contrasting manner in which films and television were conceived has resulted not only in contrasting products, but also in our different approaches to them.

The relationship between government and broadcasting is also a result of how a society conceives of the medium. Consider, as a case in point, the differing government roles in broadcasting taken in the United States, England, and continental Europe. Whether a broadcasting system is owned and financed by the government or by private corporations is the result of a cultural response to the same facts about broadcasting. The *nature* of broadcasting is no different—simply the way it is *conceptualized*. The conceptualization affects the quantity, character, and control of its products. Thus the way we perceive the new electronic media becomes a central issue to us as citizen consumers living in and helping to create a culture that directly affects our personalities.

The power of television took us by surprise. In one decade it exploded on the landscape as the ultimate consumer product, next to a house and a car, the symbol of the good life in America. It was the same decade that began with the first tentative coverage of the Republican convention that nominated Dwight D. Eisenhower and that ended with the broadcast of the Kennedy-Nixon debates, which many believe determined the outcome of the election. In one burst, television had become the central nervous system of our society, the key communications medium of our body politic—and social. In fact, it affected everything, and in economist John Kenneth Galbraith's opinion, even "the industrial system is profoundly dependent on commercial television and could not exist in its present form without it."

Television achieved its role because of its inherent technology, which makes the televised experience readily internalizable. Television can rapidly and ubiquitously become part of our individual experience and frame our social experience without the filter of self-reflection. It was these technological qualities, rather than the specifics of programming, that caused the American consumer to swallow television whole. Consequently, if there is any such thing as a national mood, a national imagination, a national outlook, it is informed, shaped, and reflected in the arena of television.

Television happened rapidly and without design. Its conception was in its consumption, and the appetite for its experience prevailed over thought. What we must explore is how the expectations of the audience and their conception of the medium limit both the style and content of what can be communicated through the medium, or leave it unencumbered to grow like a weed in the garden of our imaginations.

The Television Audience

There are two reasons for exploring television in order to understand the possible impact of the new media on our personal and social lives. The first reason is that television technology is a phenomenon of our own lifetime. We have had a direct personal experience of its explosive arrival, and we can turn to no one other than ourselves when describing what society did with that technology. Television as an historical fact is not lost in some unimaginable past as is the invention of printing. Whatever happened, we, not remote ancestors, did it. Whatever the technology's impact on society and culture, or however we describe its before and after, it all happened in and through ourselves. We did it, and we are the same people now confronted with the equally powerful and attractive new electronic media that will surely shape our future independently unless we have a constructive idea of how we intend to use them and what we want them to do.

The second reason it is worthwhile to explore television's advent here is that the new electronic media gain much of their power and wonder from the fact that they plug in to our home television sets. Thus what we have done with that television set in the past is the foundation for what we may do in the future. What we individually and collectively have come to affirm, believe, assert, suspect, and expect of a television set is precisely where we will inevitably begin when deciding which of the new media to purchase and how we will use them once they are plugged in to our television sets.

It does not take very long for social attitudes and expectations to become fixed. By the time Newton Minow, chairman of the Federal Communications Commission (FCC) during the Kennedy administration, described television as a "vast wasteland" in the early 1960s, our ideas about television were fully established. Since then, it has become commonplace for television to be accused of being nothing more than a "boob tube," an "idiot box," a mindless pacifier for children and the childish, a condition of the current illiteracy, the cause of lower moral standards, the trigger of violence and juvenile crimes, and the compelling purveyor of commercial items we neither need nor really want.

In his book *Poetry and Experience,* Archibald MacLeish complained about "the snake-lime sin of coldness-at-the-heart" characteristic of our technological civilization "in which the emotionless emotions of adolescent boys are mass produced on television screens to do our feeling for us, and a woman's longing for life is twisted, by singing commercials, into a longing for a new detergent, family size, which will keep her hands as innocent as though she had never lived." Many would agree. It seems an accurate reflection of the role television plays in our lives.

There are echoes of this criticism in Tennessee Williams's *The Glass Menagerie,* presented on CBS in 1966. There was something particularly ironic about television presenting a play built on the kind of harm that can be produced by indiscriminately mixing fantasy and reality. In the play, one of the key characters, Tom, feels trapped by his family and dreams of escaping to war or off to sea, but only finds himself at the movies. He complains that "people go to the movies instead of moving. Hollywood characters . . . have all the adventures for everybody in America while everybody in America sits in a dark room and watches them have them."

Although they sound remarkably similar, there is a significant difference between the criticism of Tennessee Williams's Tom and that of MacLeish. Most of us in the audience never blamed the movies or any particular genre of movies, or even the technology invented by Thomas Edison, for robbing us of firsthand reality. Rather we always blamed those individuals who paid the price of admission to be there in the first place and who personally substituted celluloid for the flesh and blood of real life. It was their own fault not the movies'. Quite the reverse has been true of our response to television. We have tended to blame the medium, or the technology, rather than ourselves, perceiving ourselves as victims more than actors.

It isn't as if people didn't try to warn us. From the beginning, television has been inundated with critics who bemoan their lack of influence. Several centuries ago, Goethe suggested that a critic's job was to answer three questions: What was the artist attempting to do? How well did he or she achieve the purpose? Was it worth doing in the first place? It is this third question that has gone unanswered with regard to the broadcasting of television in our society. There has been no cultural consensus, except by default, as to what the "public interest" is.

Our consistent audience behavior tends to confirm that our expectations as consumers are being adequately satisfied by the producers and suppliers of television fare. We have, in fact, created a social contract that is being fulfilled. Most of us, however, wouldn't admit to this, and most critics refuse to accept it. They simply assert that most of what we watch on television is not worth doing in the first place. Although a large part of the audience claims to agree, their behavior hasn't changed. The phenomenon of television—what we mean when we say, "I watched television last night"—has been shaped by what we, the audience, find acceptable and watch week after week, thus legitimizing it by our time and attention.

Marshall McLuhan was correct when he said that the medium was the message, and television quickly became our society's central medium of communication. It grew like a weed without the shaping impact of a consciously defined social contract to determine what the medium should be delivering, for what purpose, and with what value. Perhaps it happened too fast, too unexpectedly for constructive reflection; so that the technology shaped us before we could shape it. Or perhaps we didn't care, or know where to look, or were simply afraid to tamper with a medium of communications because as Americans we believed so strongly in free enterprise and free speech. Whatever the case, we would do well to consider from our current position the process of television's impact on our lives. How we thought about the technology, what we did or did not do about its presence in our midst, and how it became part of our interior lives suggest an outline of reflection for approaching the new electronic media. With luck, such reflection may alert us to what may or may not work in the future.

Sometime in the 1960s we gave up, as a community, trying to define what we meant by the FCC regulation that a licensee must operate a television station in "the public interest, convenience, and necessity" as it applied to television. This regulatory concept was inherited from legislation controlling radio broadcasting, which in turn had borrowed the concept from the regulation of public utilities and railroads. It seemed to make sense in the 1920s that the idea of "public interest, convenience, and necessity" could be applied to broadcasting because radio waves were analogous to the public highways and other rights-of-way developed and utilized by privately owned monopoly franchises in transportation and public utilities. In other words, the public "owned" the air just as it "owned" the land and public highways. Although it often proved difficult to lump the distribution of radio programming in the same category as trains running on time and providing sufficient gas for heat in winter, the application of the concept to radio broadcasting worked reasonably well. It worked well precisely because radio always remained on the periphery of our lives. It never became the central nervous system of our society, dominating and shaping our major institutions and behavior. Television was different. It had the power to eclipse and encompass other institutions, and it did.

By becoming, almost overnight, the dominating and encompassing medium of our society, television broadcasting functioned in a way quite different from what its makers and users had intended. Had we thought of television as having a reality of its own apart from the programs it broadcast, we could have understood television's relationship to the public interest more clearly. We could have discovered the objective social functions of broadcasting and used them to inform our choices and shape our behavior. A clearer understanding of our own fundamental community expectations would have enabled us better to judge the influence of television on society and to channel that influence in our own and the public's best interest.

Public Interest and Public Expectations

The basic social functions of communications as articulated by most sociologists of communications include:

1. Surveillance of the environment, reporting on dangers and opportunities to the individual and community.

2. Correlating of the components of society in arriving at a response to these reports.

3. Transmission of the culture, or social inheritance, to new members of the community.

4. Entertainment of the people for their own enhancement and enjoyment of life.

These are very nearly the same the functions that ancient rhetoricians gave to communications: to inform, to persuade, and to entertain.

From sprawling urban conglomerates to self-contained rural villages, communications determine the common ideas, aspirations, traditions, political action, and economic progress of the community. People have always required and utilized something to watch over their environment and report on dangers and opportunities. Societies have always needed something to disseminate facts and opinions, to help citizens make decisions and then circulate the decisions throughout the community. Families have always needed help in passing on the lore, wisdom, and expectations of society to the new members of their community. Each society has always needed a forum to entertain its people in a community experience, to give individuals an opportunity to relax, refresh themselves, and enjoy their lives. Finally, communities have always needed something to broaden their trade and commerce, a medium of buying and selling.

Small communities can perform these functions through face-to-face communication—at home, school, church, and marketplace. As communities grow larger, however, more extensive communications media are devised, which in turn allow for even more growth. In the 1950s television quickly took over these social functions which were previously performed in face-to-face communications and through the multiple small-scale media of newspapers, magazines, and radio. None of this was a secret to scholars. They witnessed it occurring in the very decade in which television exploded on the scene. What was about to happen—and *was* happening—was knowable. Professionals in the industry and the leading social thinkers of the day could have educated the American public about the need for a carefully defined social contract regarding the role of television in our society. The four objective functions of communications could have been adopted as norms for an effective social contract between broadcasters and their audience of citizen consumers. The technology cried out for such a contract precisely because it had the ability to eclipse and preempt all other social media. It had the power to be singular!

Had we been more aware and more articulate about the social functions that television was clearly assuming, we might have achieved a consensus on four norms for programming and viewing.

1. Television should contribute to the quantity, depth, and complexity of the flow of current information in a manner proportionate to the concerns of society and the commitments we make.

2. Television should present the analysis of information and debate of public issues in a manner practically sufficient for the proper exercise of democracy.

3. Television programming should offer a true picture of reality and reflect society's cultural values.

4. Television programming should enhance our experience and help us to live abundant, sensitive human lives.

These four principles could have formed the foundation for our social contract and defined our expectations and given substance to what we meant by the "public interest."

The focus here is not on breast-beating over the shabby products of broadcast television. Cultural elitists constantly attack the programming we watch on television while insisting that other systems of broadcasting, particularly England's BBC, are strikingly superior to our own. On the contrary, the American broadcasting system is by far the best in the world, and those who celebrate other systems from afar undoubtedly have had little firsthand experience of watching television twenty-four hours a day for a week or a month in these other countries. While acknowledging that many American programs contain nothing but palpable pap, we should not overlook the programs that successfully offer content of the highest quality. Nor should we forget that when evaluating communications, as when evaluating libraries and museums, quantity is one of the characteristics of quality, and the American system is far ahead of other countries in the sheer quantity offered to viewers.

But in considering these four norms of programming and viewing, it is clear that broadcast television escaped critical evaluation on the individual and collective levels. Since we as a society failed to articulate our hopes for television based on its objective functions and as a result failed to translate these hopes into workable expectations, we gave up practical control over a medium that changed our personal lives and personalities. Neither in our private hours of watching television nor in our public discussion of it in society at large were we able to moderate and direct the impact it would have on our culture. There was no positive contract between the sender and the receiver of television programming. There was no moderating self-consciousness about how the technology and the culture would come to terms with one another.

Modern Americans, brought up on the story of how the Battle of New Orleans was fought after the treaty ending the War of 1812 had been signed because it took so long for the news to cross the ocean, can feel smugly self-satisfied that news and public discussion are now adequately covered by the technology of communications. In these days of electronic journalism, with reporters in every capital of the world, we absorb international news events with a sense of superiority, confident that the world as we need to know it is coextensive with the world as we do know it through television. We are treated to "living room wars" from around the world.

Yet more and more, the world that we have to deal with politically, economically, socially, and morally is out of personal sight, and our individual decisions are based on facts we cannot see, touch, smell, or hear for ourselves. Facts reach us, not in the form of personal perception, not as something we ourselves have seen or heard, but in the form of communications, as something we have only heard *about*. We have come to rely on television communications to perform the same function for us that perception does in our personal lives. In the last twenty-four hours, 83 percent of our population watched a television news program, using it as their primary plug into the events of their city, state, region, nation, and world. In fact, all of us, in one degree or another, rely on television for our knowledge about the world, and if we haven't "seen it on television," it isn't real—it's as if the event has not really occurred.

Critics have long complained about the thinness of television news, comparing it unfavorably to our leading newspapers. But the essence of the problem, if one exists, is not shallow reporting. It is that we have formed a social contract by default without establishing an equivalent and demanding set of expectations. The question is: Does television provide sufficient information to allow the citizen to participate rationally in the vital activities pursued by society? Is there, for example, any sense of proportion between the impact of our votes and the information and understanding on which they were based? If we answer no, we have to ask why, and for that answer we need not look beyond ourselves.

The key fact of our communications life is that the level of citizen expectation is a key component in a democratic society shaped by television technology. It determines the citizen's level of attention on the one hand, and the quantity and quality of analysis and debate of public issues provided by broadcasting on the other hand. It is for reasons such as these that federal regulators continue to insist on the maintenance of the "fairness doctrine," "equal time," and the availability of commercial time for political parties. But such regulations are not the equivalent of positive expectations and demands. They are not a substitute for aggressive management decision on the part of broadcasters. Nor are they a substitute for active citizen interest.

We have learned how to do the news better, and Nielsen ratings have supported the increased time devoted to news in a broadcast schedule. All is not wrong with America. But these changes have occurred with a distinct lack of self-consciousness on our part as citizens. While broadcasters probably follow or reflect public taste more often than not, they are simultaneously creating our social perceptions. Consequently, we are living with all the conditions of a social contract without any prior conceptualization of what we are doing, or wish we were doing.

When it comes to entertainment programming, on the other hand, many parents and critics turn into social engineers, anxious that the medium be used to describe and preserve their particular view of the world, or as an instrument for promoting their personal social ethics and etiquette. Their arguments center on the vacuity of children's cartoons, the psychological impact of a steady diet of violence, the sexual premises, innuendos, and cavortings of prime-time comedies and dramas, and the failure of series programs to represent a cross section of society, while stereotyping the people, roles, and professions that are depicted. Generally, these parents and critics insist that someone, either the networks or the government, do something at the head end of the supply system to determine our imaginative diet. It is a manipulative response to the fear of being manipulated.

The tragedy of this approach is not only its ineffectiveness, but also its appeal to an authority outside ourselves—thus creating a vacuum of personal responsibility. The public arts, from Athenian theater to American television, succeed or fail in their ability to capture and maintain public attention. It is part of the theatrical art to "put fannies in seats," and failure to do so is an artistic failure, just as the inability to obtain votes and consensus is a political failure. The burden for determining successful programming in style and content is on the viewer, just as much as political results are a citizen responsibility. If people do not insist on, and refuse to watch anything but, programs that strive to present a true picture of reality, that enhance their experience and enlarge their vision of human worth, then the programming menu will continue to follow the tastes viewers do express by their acquiescence.

The broadcast networks have often argued that the citizens determine viewing choices by their viewing habits, which are accurately measured by the Nielsen ratings. They have a point. Armed with an on/off switch and a channel selector, the citizen votes on broadcast programming. By deciding what it will and will not watch, the audience legitimizes a social contract of expectation and satisfaction established between sender and receiver, supplier and consumer. And all of us are responsible for being receivers and consumers—as attested to by advertisers who can and do reach nearly 100 percent of the adult population with their commercials judiciously spread throughout the three network schedules. In other words, everyone watches television. A social contract is being acted out, and the programming we choose to watch tells us who we are, whether we like it or not.

A Technology Responsible to No One

Who we are, however, is people who didn't choose to think ahead about the shaping influence of television on our social character. In our public discussions of the public interest, as well as in our private behavior, we demonstrated a weakness in understanding, insistence, and personal discipline. Television rapidly became our primary medium of information, discussion, entertainment, and self-reflection, and consequently, it has determined to a great extent our social and cultural values.

But without a conceptualization of television value and expectation—what is worth doing and watching in the first place—and its importance in society, we, the people, just watched "television"—which necessarily became responsible to no one.

That is the history of our collective behavior in the face of the last explosion of communications technology, and we are the same people faced with an equally powerful explosion of new technologies. What makes this situation so striking, however, is that it was possible to understand television's role in our society at the time it was advancing, and it could have been possible to establish a corresponding set of social objectives and personal behavior to direct that advance. Our failure to act was both a failure in imagination and a failure of will.

Over the years, most of the political, regulatory, and social debate over broadcasting has focused on the unique condition of broadcasting—that it utilizes the airwaves, which are owned by the public. Consequently, the FCC long ago established as the terms of the debate that "the paramount and controlling consideration [is] the relationship between the American system of broadcasting carried on through a large number of private licensees upon who devolves the responsibility for the selection and presentation of program material, and the Congressional mandate that this licensee responsibility is to be exercised in the interests of, and as a trustee for the public at large which retains ultimate control over the channels of radio and television communication."

All sides of the argument have always agreed with the FCC that "the interests of the listening public are paramount and may not be subordinated to the interests of the station licensee." A difficulty is created, however, by the fact that these licensees are private corporations organized to make a profit. The Supreme Court once made it very clear that "a business corporation is organized and carried on primarily for the profit of the stockholders. The powers of the directors are to be employed for this end."

The problem of a private corporation becoming the custodian

of the public interest has always alarmed both moral conservatives and the liberal cultural elite. Even Adam Smith, the great spokesman of the free enterprise system, once wrote that any attempt to regulate commerce by laws devised by business interests themselves is always suspect. He argued that such regulations come from men "whose interest is never the same with that of the public, who have generally an interest to deceive and even to oppress the public, and who accordingly have upon many occasions both deceived and oppressed it." Other countries have solved this problem by creating government-owned broadcasting channels, an arrangement that would radically alter American society.

Discussions over the public interest versus privately owned broadcast channels most frequently end up, however, with a demand for the strict regulation of both program content and style, based on someone's view of what is good and proper. Another response, of course, is that the ratings prove that the broadcast stations are providing precisely what the people want, so there is no need to tamper with it. Presumably the people's interests are being served. But throughout the discussion, the key counter arguments constantly and consistently revert to the issues of free enterprise and free speech—for free speech requires free enterprise and free enterprise requires free speech—which mutual dependence forms the very fabric of American culture. And so, the debate will go on forever. The focus of our discussion, therefore, is not on specific laws and regulations but on "we, the people," and our expectations and behavior.

The Failure of Public Broadcasting

Did the public ever have an honest opportunity to vote for quality programming? Those who argue that it did not, note that nearly 40 percent of our citizen homes, during any given prime-time quarter-hour, do not have their television sets turned on. The assumption is that this 40 percent were not given the opportunity of expressing their votes; and that, had they been offered quality programs, they would have been active participants and the base audience for such programming in the future. If that were so, then everyone would have had the opportunity to be attracted to a higher standard of news and entertainment.

In the late 1950s and early 1960s, however, it was considered axiomatic that advertisers, the sole support of our commercial broadcasting system, would not support such programming alternatives, which thus would never surface within the network schedules. This was, in fact, a key argument behind the founding of the Public Broadcasting System. Another motivation for the creation of public broadcasting was the widespread perception, once articulated by Edward R. Murrow, that "if television and radio are to be used to entertain all of the people all of the time, then we have come perilously close to discovering the real opiate of the people."

In 1967 the Carnegie Commission on Educational Television, whose report and recommendations led to the establishment of public broadcasting, viewed the sociocultural opportunity this way:

> We have become aware of television as a technology of immense power, growing steadily more powerful. What confronts our society is the obligation to bring that technology into the full service of man, so that its power to move image and sound is consistently coupled with a power to move mind and spirit. Television should enable us not only to see and hear more vividly, but to understand more deeply.

The Commission imagined a continuum of programming that "replenishes our store of information, stimulates our perceptions, challenges our standards and affects our judgment." It saw television and life as inextricably united, and both in constant process of educating and forming the American mind.

In the end, the commission recommended the construction of a programming system funded at public expense. It was their judgment that only through government support could broadcasting "show us domains of learning, emotion, and doing, examples of skill, human expressiveness, and physical phenomena that might otherwise be outside our ken," and "lift our sights, providing us with relaxation and recreation, and bring before us glimpses of greatness." In their view, public television would include "all that is of human interest and importance which is not at the moment appropriate or available for support by advertising." In short, public television would include anything that wasn't commercial.

After all was said and done, advertising was the culprit. Not the technology. Not the imaginations and expectations of citizen consumers. Not the nature of broadcasting or the social contract between programmers and public. But advertising.

It was never advertisers, however, who stood in the way of providing what was then called "alternative programming"; rather it was the nature of the broadcast technology itself. It is a fundamental misunderstanding of advertising to believe that advertisers value all consumers and all programming equally. Advertisers are constantly struggling with the art and science of their profession—the art that alerts them to media environments they ought to be in, and the science that measures the size and demographic characteristics of particular media vehicles. In any case, it is the rare advertiser who is interested in every- and anybody, every- and anywhere, at every and any time. Many advertisers would willingly pay premium prices to advertise on programs designed for and watched by relatively few people with specific interests and characteristics that match their product's or corporation's personality. They "underwrite" so many public broadcasting programs precisely in order to reach these prime prospects with their message. In fact, we call those few early years when advertisers could influence programming, "The Golden Age of Television." Since the 1960s, however, advertisers alone do not have the ability to leverage such programs onto the network schedules.

The brutal economics of supply and demand that govern television are rooted in its technology. Whether commercially supported or not, broadcasters are interested in reaching the greatest number of people simultaneously and continuously available to their programs. Why? Because that is what the technology can do. You do not use a technology that can light up the whole sky to search for a cat in an alley; that is a task for the technology of the flashlight. As a technology, broadcasting can reach everyone simultaneously, and any artist or business person, any editor or publisher, or politician, using the technology inherently wants to communicate with everybody all at once and all the time. Such an aesthetic is rooted in the technology, and you cannot blame a practitioner for trying to achieve it. That is the motivation of ABC, CBS, and NBC, and should have been at least part the motivation of PBS.

Advertisers are on commercial television because business is always conducted where people are communicating with one another. The audience is composed of citizen consumers who always hold the power, and on a massive scale they watch network television. We as citizens and consumers are the beginning and end of the process, and that is why our concept of the medium is the foundation of the social contract, as well as the economics of the system.

The creators of public broadcasting tried to escape both reality and technology. They completely overlooked the fundamental interrelation of free enterprise and free speech in America. They attacked advertising, perceiving broadcasters only as the creators of an advertising vehicle. They completely overlooked opportunities for the "public interest" to be articulated in a free-enterprise communications environment, and how well commercial broadcasting performed. They felt it necessary to go outside free enterprise—which is to go outside our culture.

In trying to escape American culture, public broadcasters were also trying to escape the nature of the public arts, which have always been swarthy, sweaty, sexy, and spectacular in their drive for popularity, for popularity is essential to their art.

In fact, public broadcasting has tried to impose an elite's standards of taste and relevance on a diverse and democratic population. It didn't and couldn't work. By turning away from the necessary limitations of broadcasting and the public arts, and by focusing on a government-sponsored solution, they created both an aristocratic cultural chasm and an economic stepchild.

If we had had an appropriate conception of broadcasting as both a product and a creator of our society, there would have been no need for a "solution." Furthermore, without such an understanding, the proposed solution went outside the continuum of society communicating by injecting a government-sponsored alternative culture. Because public broadcasting was not at the heart of our culture, as commercial broadcasting was, it necessarily lacked the broad public support necessary for its adequate funding.

"Sesame Street" and "Masterpiece Theater," ballets, symphonies, and documentaries may be what broadcasting should be all about; and I, for one, am not unhappy with their availability no matter how they arrive in my home. None of this should distract us, however, from the fact that public broadcasting doesn't work as an institution, or that its failure is rooted in its conceptual foundations.

Now that public broadcasting is experimenting with advertising, it may appear to be Monday-morning quarterbacking to question our earlier conceptualizations. But the key issues relating to broadcasting and the public arts were known in the late '60s. More important, the irrelevance of broadcast technology to what turned out to be the actual goals of public broadcasting was known at the time.

In a paper submitted in 1967 to the Carnegie Commission and published as a supplementary paper to its final report, Dr. J. C. R. Licklider pointed out that the great characteristic of broadcast television is that it is *broadcast*. He argued that "to justify the use of a medium capable of carrying a very large amount of information each second, one had to reach a mass audience" and that this was the "*sine qua non* for a wide-band medium." He recognized that "from an educator's point of view, the elite would feel left out." Yet one could not escape the intrinsic characteristics of the technology. The "conflict was between the effort to select material of interest to the individual and the commitment to broadcast to a mass audience." The technology challenged the objective and thwarted its achievement.

What the founders of public broadcasting were looking for was "narrowcasting." Dr. Licklider claimed to have coined this buzzword of our time. He wrote:

> Here I should like to coin the term "narrowcasting," using it to emphasize the rejection or dissolution of the characteristics imposed by commitment to a monolithic mass-appeal, broadcast approach. Narrowcasting may suggest more efficient procedures than broadcasting throughout a wide area in order to reach a small, select audience, and it is meant to imply *not only that the subject matter is designed to appeal to selected groups, but also that the distribution channels are so arranged* as to carry each program or service to its proper audience. [Emphasis added.]

Not only was broadcasting not the right technology, but its misapplication indicated a misconception of the technology itself. Licklider's paper is evidence that this was known at the time, and he was not alone. The National Association of Educational Broadcasters published a report in the late sixties, pointing out that not only were the new technologies of cable television, videocassettes, and videodiscs more appropriate to the tasks of diversity, but that they would probably be economically viable in performing this function. Consequently they argued that the arrival and pervasiveness of the new electronic media would undermine the financial support, and probably the legislative mandate, for public broadcasting. Today, with the Arts & Entertainment Network, C-Span, Bravo, the Disney Channel, Nickelodeon, and CNN all available to cable sub-

scribers, and ballets and symphonies available on videocassettes, public broadcasting is more apparently unrealistic at its
base.

The history of public broadcasting teaches us that it is useless
to go against a society's historical conception of a medium, for
in the end, public conception will determine public communications. This history also teaches us something about applying
the right technology to the right goals. Technology will always
rebel against its misapplication in either economic or artistic
failure, and then gain its revenge by developing new and more
appropriate technologies. If the Carnegie Commission wanted
to reach small audiences, broadcasting was not the way. There
is a technological lesson to be learned in all this, but it should
not eclipse the human lesson. It was still *the audience's use* of the
medium that determined its outcome.

McLuhan's Message

Television has always had an epistemological dimension for
all of us. It has always been a means of knowing and experiencing, as much as it has been a medium of entertainment and
advertising. How people used the medium and how it shaped us
as we consumed it was definitively influenced by Marshall
McLuhan, the accepted guru of television as a field of knowledge and means of knowing. In his two key works, *The Gutenberg Galaxy* and *Understanding Media,* he focused on the role of
communications technologies in society. His vision that the
"medium is the message" was conceptually valid, although like
many gurus he was probably extreme in his applications. The
medium is the message, McLuhan argued, because the dominant medium's way of knowing, perceiving, and expressing
determines a society's self-consciousness and behavior.

McLuhan proposed that each medium—especially printing
and television—has a specific grammar and vocabulary that
establishes prior categories of judgment and response. The par-

ticular media of a period affect the human spirit in particular ways. Thus the linear, sequential, one-thing-after-another characteristic of printing imposed a linear, sequential, one-thing-after-another gestalt on society's behavior and worldview. In McLuhan's view, these characteristics of the printed page created the mind-set for separating work into specialized functions in an industrialized mass society.

At a time when many people were afraid of becoming anonymous ciphers in corporate life, McLuhan highlighted their fears. He contrasted "civilized man [who] tends to restrict and enclose space and to separate functions" with tribal man, who "had freely extended the forms of his body to include the universe." Television was the corrective, the ameliorating technology, which could return us to a unified understanding of ourselves and the universe by establishing "a global network that . . . constitutes a single unified field of experience." McLuhan believed that television would involve all people deeply in one another and that in the process we would retrieve ourselves from the anxieties of functional separation and superficial homogeneity of mass society; it was television that would foster "uniqueness and diversity."

In McLuhan's vocabulary, television is a "cool" medium, necessarily involving more than one sense. Television images have "low intensity of definition," and unlike other media do "not afford detailed information about objects." Since the television image is a mosaic mesh of light and dark spots conveying low-information quality, it invites "participation or completion by the audience." Television is a "do it yourself" medium.

McLuhan believed that television must be "felt through" by the viewer who must "unconsciously reconfigure the dots" in order to derive meaning from them. In contrast to print, which is an intellectual and passive medium, television is at once intellectual and participative. The audience, he argued, "participates in the creative process." Consequently television "fosters many preferences that are quite at variance with literate uniformity and repeatability." He concluded that "our entire standardized economy" may despair over this, but due to the pervasiveness of television "the uniform and repeatable must now yield to the uniquely askew."

What most college students, teachers, and the general public took from McLuhan was that technology dictates society and that television was the solution to the angst of mass culture. In McLuhan's view, the technology of television induces individual and personal involvement. The "individual had technological control" at the reception point and, through a process of free association of images of experience, entered into a coextensive relationship with the world. The viewer becomes a citizen of a "global village"—a situation that paradoxically encourages individuality, precisely because each individual uniquely determines the patterns of the images he or she experiences, using television as a personal tactile reach into external life.

To the extent that we became McLuhan groupies in the 1960s, we accepted the idea that television would change society for the better because televised experiences, alone and together, could not mean what they said, nor say what they mean until freely given meaning by the viewer. We embraced an epistemology based on self-created logic and an undifferentiated celebration of individual perception. To the extent that this idea affected our actions, objective logic and universal significance evaporated from our lives.

In addition to enjoying television's reduction of thousands of words to pictures, we used the technology to foster a rampant subjectivism. We used television to substitute perception for cognition and emotion for thought. We allowed the power of its images to motivate and inspire us to act without planning or thinking through the consequences of our actions. Television, in our McLuhanesque hands, became the almost ideal medium of process-oriented lives. Bereft of objective goals and ridden with corporate angst, our citizen consumers not only embraced the then-new technology of television, but saw in it a vital new epistemology for their subjective lives.

Experience for the Future

The victory of television technology was that it was unrestrained, not in the sense of government regulation but by the absence of an appropriate social contract of expectation and satisfaction based on a functional understanding of the medium. No medium before television came upon us so fast, or was equally uninhibited by time, place, or culture. At the time, we were all too busy "getting and spending," in communications terms, to find time for contemplation to measure action. Thus, although the scholarship was available for grasping the functions of television as the central mass medium of our society, and for translating those functions into the shared expectations of a social contract, we overlooked them.

The history of audience indifference and limited funding that has plagued public broadcasting should alert us in the future that the public arts are necessarily democratic, and you cannot effectively utilize a mass audience technology to serve an elite's objectives. The grounds of every technology, like the grounds of every art, are knowable and definitive. In many ways technology is an art form, and not to know its form is to fail at the craft, to fail at knowing what it can and cannot do.

The effective use of television that touched our personal, interior lives was its ability to liberate us from social and cultural constraints, to encourage us to formulate our own concepts of fact and value from television's images of experience. McLuhan's epistemology is the most dramatic and effective articulation of this response. It supported a technology run rampant precisely at its point of reception.

The point of reception was ourselves. Our primary concern, after all else is said and done, is our own lives, personalities, and character. Television affected all three. In the next chapter, we will examine our social character as it relates to communications technology and discover a possible approach for addressing the future.

2

The Media and Social Character

In 1948 Yale sociologist Harold Lasswell published a paper in which he formulated what has become the classic definition of communications. He said that "a convenient way to describe an act of communication is to answer the following questions: 'Who—says what—in which channel—to whom—with what effect?'" (Of course, to the extent that one accepts the proposition that "the medium is the message," the question of who says what to whom is correspondingly irrelevant, because the channel of communications, the medium or the technology, will decide the outcome.) In dealing with television, most of us—audience and critics alike—have consistently overlooked the "To Whom" aspect of Lasswell's definition. We have focused on the networks and their programming, or on the technology referred to as "the power of television." But we seldom looked closely at ourselves. The primary questions of the future, however, are, Who will we be, we who use these new communications technologies? And how will using them change us? In short, what new social character will emerge from our using them?

To answer these questions let's turn again to the history of television, which provides an almost laboratory opportunity to think of social character and technology both separately and together and to isolate their effects on each other at a very personal level. The fifteen postwar years during which television suddenly occurred were a rather pure moment of communications history and a rather precise moment of social history. Almost four hundred years intervened between the printing of the Gutenberg Bible and the circulation of the New York Sun, the "penny-post" that opened the way in 1833 for mass-circulation newspapers. Almost another hundred years intervened before the first radio networks were established during the tumultuous times of depression and war. Thus, although we can track the impact of these mass media, it is difficult precisely to isolate their "before" and "after" at cohesive, condensed moments of social history. With television, on the other hand, we have a rare opportunity to identify a formula for confidently projecting the effects the new electronic media may have in our lives.

The Media in a Mass Society

Our general inability to discover how television functions on a personal level is the result of our thinking in terms of a mechanistic, stimulus-response model for mass media. It has been easy to imagine the concentrated power of the networks radiating their messages on the unsuspecting masses, and convenient to assume that the masses are being forced to wallow in the mire of mediocrity by the purveyors of mass products. This is the facile picture of television in our heads. Is it justified?

This concept of mass communications—holding mass society together and giving it direction for good or ill—dominated communications theory into the 1940s and persists even today. From this point of view, advertising is irresistible propaganda for unwanted consumption; television programming directly results in social violence and decaying sexual morality; and political journalism is perceived as political control.

Psychoanalytic theory, notably Freudian theory, supported a direct stimulus-response theory of communications, suggesting that each person responded in an almost involuntary way because of a common psychic structure. Consequently, "hidden persuaders" could launch subtle campaigns against the unsuspecting masses, compelling them to buy whatever commercial products were being sold that season. Advances in research methodology since the 1940s, however, have led to a very different view. New insights into demographics and psychographics suggest that an audience is composed of social groups and personality groups, rather than isolated, yet homogeneous, individuals. In general, these studies moved away from the more mechanistic stimulus-response theories of communications with their emphasis on "Who—says—what" to a focus on the audience and the role the receiver plays in changing the communication and determining its effects.

What the new empirical data indicated was that variables such as your age, where you lived, your sex, income, and education made a substantive difference in the effects of the same communication. Content did not signify the same thing to everybody. It was only after this discovery that communicators, especially politicians and advertisers, but also editors and publishers, began to target their communications to specific demographic groupings and to modify their content accordingly. Researchers in psychology, particularly in the areas of learning theory, behavior modification, and conditioning, discovered that our individual personalities, attitudes, values, and beliefs systematically altered our perceptions of reality. Consequently, different people not only select different messages from the media in general, but also select different stimuli within the same message. They not only come to different conclusions, but they have different experiences. Hence communicators and editors used this information to elicit a desired action by reinforcing attitudes already present in the listener. In this light, any communicated message, because it is not operating in a vacuum, is clearly more an agent of reinforcement than of change. So also is television programming.

Television may have monopolized our lives, but it wasn't a monopoly. Television competes with all other media for our time, attention, and leisure activities. Furthermore, television is intensely competitive within itself. Each member of the television audience picks and chooses among available programs and individually determines their significance. The inescapable conclusions are that the content of communications is *not* dictatorial, and that if mass media hold us together as a society, they do it in a more complex way than we previously supposed.

Social Character

The product of society communicating is culture, and the product of culture is social character. Social character is the point where culture and personality meet. It is the effect of communications. Hence our current social character results from the dynamic of the dialogue inherent in our current communications technology; broadcast television very rapidly became the key medium of our society communicating.

As Christopher Lasch has pointed out in *The Culture of Narcissism,* "every society reproduces its culture—its norms, its underlying assumptions, its modes of organizing experience—in the individual in the form of personality." The result is social character. In *The Lonely Crowd* David Riesman described social character as "the more or less permanent socially and historically conditioned organization of an individual's drives and satisfactions—the kind of 'set' with which he approaches the world and people." Social character, says Riesman, is the shared perspective of social groups in response to their changing environment, and "the way in which society ensures some degree of conformity from the individuals who make it up." In mapping the key effects of television we must discover its specific role in the development of our social character. Such an effort has more than historical importance. It could be prophetic. After all, the bottom line for the future depends upon the shared personality we develop as new technologies enter our society.

David Riesman wrote *The Lonely Crowd,* with co-authors Nathan Glazer and Reuel Denney, in the late 1940s and it was first published in 1950. In it he described the social character of America at the dawn of the television era. To contemporary readers who cannot imagine a world without television, it is remarkable that Riesman mentioned television only twice and then just in passing. His portrait of the American character is as it was without television, just before television suddenly happened, an American character unable to imagine or expect the impact of the new technology at its doorstep.

This apparent accident of chronology could be reason enough for us to pay particular attention to *The Lonely Crowd,* but, as a somewhat baffled Riesman observed twenty years later, "*The Lonely Crowd* has in some measure entered the picture many Americans have of ourselves both past and present." Not only do scholars keep coming back to his weltanschauung, but the enthusiastic acceptance by the general public was like that of a generation to its own definitive novel. Thus his work is probably as close as we can get to what most of us were like on the day before we purchased our own first television set.

Riesman was primarily concerned with the "mode of conformity" produced first by the industrial revolution and then its aftermath, particularly the "whole range of social development associated with the shift from an age of production to an age of consumption." He used the terms *inner-directed* and *other-directed* to describe the social characters that emerged from this revolution. Before the industrial revolution and the increase in literacy it brought about, societies were relatively static, and individuals tended to reflect the patterns of the age-grade, clan, or caste of which they were members. Older relatives taught the younger generation everything they needed to know, social mores endured for generations, and individuals rarely traveled far from their roots. Literacy and industrialization shattered this type of conformity, only to produce another.

In Riesman's view, the spread of literacy and industrialization was rapid and dramatic. Suddenly, there was mobility and expansion everywhere, an eruption of communications, knowledge, and literacy, an intensification of economic opportunity, and a radicalization of individual choice. Society was confronted with a need to secure social conformity when circumstances demanded initiative and independence.

The solution that evolved was a new social character in which, according to Riesman, "the source of direction for the individual is 'inner' in the sense that it is implanted early in life by the elders, and directed toward generalized but nonetheless inescapably destined goals." In Riesman's terms, the inner-directed person incorporates a "psychic gyroscope," set going by his parents, that "can receive signals later on from other authorities who resemble his parents. He goes through life less independent than he seems, obeying this internal piloting." The cost of straying off course is a feeling of guilt. Many people have identified rationalized, individualistic attitudes controlled by a strict morality, and supportive of the accumulation of wealth as the "Protestant work ethic." Whether the Protestant ethic is perceived as coextensive with inner-direction or as an example of it, inner-direction was useful in harnessing the driving forces of individualism with moral imperatives.

As society changed once again, moving from an age of industrial production to an age of consumption, the key to our daily work became the move away from physical manipulation toward conceptual manipulation, from working with things to working with people. These changes both demanded and created a new social character, which Riesman termed *other-directed*. For other-directed people, their contemporaries are the source of direction. The goals and values that motivate the other-directed person shift as easily as his contemporaries shift; only the *process* of paying close attention to the signals from others remains unaltered throughout life.

What is internalized is not a code of behavior, but the elaborate equipment needed to attend to messages from others about behavior. The other-directed person tends to become merely the succession of roles he or she plays. No clear line exists between adjusting to the group and serving one's own private interests, and the prime psychological lever of the other-directed is a diffuse anxiety, rather than guilt. Essentially, this adaptation required the rechanneling of inner-directed competitive drives into other-directed drives for approval. For an internalized gyroscope, the other-directed substituted an internalized radar.

In developing a typology of inner-directed and other-directed to describe the two social characters, Riesman admits that he "focused on changes that most readers seem to have regarded as changes for the worse." For example, when William Whyte gave a generational brand name to other-direction with the publication of *The Organization Man,* he warned against the transformation of American society into a corporate ethic of security seekers. He worried about an emerging middle class of middle managers and junior executives who have left home "spiritually as well as physically, to take the vows of organization life," and to become the "mind and soul of our great self-perpetuating institutions." Whyte mourned the decline of the Protestant ethic, and fed the nostalgia of all those middle managers who preferred to see themselves as independent, self-reliant, entrepreneurial, and individualistic while wearing their gray flannel uniforms.

Such nostalgia leads us to overlook the fact that for the vast majority of American workers, those laboring in coal mines, steel mills, and automotive plants, the only way to improve their condition was to improve the condition of their group. The self-reliance of self-made entrepreneurs was as inaccessible to them as to today's migrant farm workers. Inner-direction characterized the social character of the age, but for the majority of workers it was a discipline of conformity.

As a survival tactic, inner-direction was incapable of offering much resistance to other-direction. Inner-direction was a type of conformity without a personalized belief system; it offered goals that needed no serious reflection or critical evaluation. Inner-direction meant Victorian morality, a class-conscious code of social etiquette, heroic belief in self-reliant individualism, and at least external acceptance that income and self-respect were somehow coextensive. Inner-direction was an instrument of controlled experience expressed as conformity, and the conformity operated as a morality for the group, a matrix of beliefs without meaning. Once perceived as the trading-in of one conformity for another, other-direction is a far less disturbing phenomenon.

Daniel Yankelovich is particularly articulate on what it was like for the majority of Americans in the postwar years. In *New Rules: Searching for Self-Fulfillment in a World Turned Upside Down,* he points out that between 1950 and 1973 the average family income doubled—from $5,000 to $12,000 in constant dollars—moving the mass of Americans from the edge of poverty to modest material comfort. As Yankelovich describes it, "If I worked hard, observed the rules and learned to keep my personal desires mostly suppressed, I might find myself well rewarded—with moral self esteem for my self-denial, with the acceptance of others for my respectability, and with worldly goods from an affluent economy." For the vast majority of Americans for whom inner-direction was a survival tactic, the new rules of other-direction, if that's what they were, paid off.

There was a cohesiveness to American society then; and success, in terms of a good home in a good neighborhood, sending your children to college, and traveling on vacation, was available to millions of people, the masses, who before the war could only dream of such things. They did not denounce consumer society for its materialism, but rather perceived consumer expectations as common to all individuals. These citizens vigorously invested in durable goods, which they used for years as the foundation of their delightfully defiant and exuberantly new life-style. It was the fulfillment and democratization of the

American Dream. If in the process they traded in their inner-directed gyroscopes for other-directed radar scanners, they certainly can be understood and perhaps forgiven.

During the rush to affluence that characterized the postwar years, four of the key agents of character formation changed. Work, school, family, and the peer group exerted new and different pressures in the process of society communicating.

In the postwar years, work was more and more an agent and a symbol of one's life-style than an instrument of survival. In the exchange, the typical worker both limited and surrendered his skills to the needs of the corporation. Work became white-collar labor for wages. This resulted in two things. First, because of a change in the nature of how and with what we worked, group opinion intensified in value. Second, the opportunities for competition and the tools to compete changed. Very few individuals, if any, persisted in the dream of making their own fortune by their own hand or efforts. Horatio Alger had always been a myth for most working people, but suddenly it was a myth for the college-educated, whether armed with the Protestant ethic or not. Christopher Lasch notes in *The Culture of Narcissism* that the only thing they could achieve or make in corporate life was success understood as winning. Lasch calls it "a form of sibling rivalry, in which men of approximately equal abilities jostled against each other in competition for a limited number of places." Thus the typical worker became, in Erich Fromm's phrase, "a marketer of oneself."

This shift in the nature of work, in both the opportunity and means of affluence, produced significant changes in education. Fundamentally, education shifted from a curriculum of knowledge to a curriculum of opinion, training the population in the technical skills of other-direction. Riesman suggested that as "people have become more other-directed, educational methods tend to thwart individuality," and the teacher "conveys to the children that what matters is not their industry or learning as such, but their adjustment in the group, their cooperation, their (carefully stylized and limited) initiative and leadership."

The goals of education evolved into developing compatible citizens, making students happy in their social lives, and creating ideal workers, citizens, and family members for the society that prevailed. Riesman saw this educational process as both supportive and derivative of our developing other-directed social character. He emphasized that the cooperation and leadership inculcated in and expected of children was "frequently contentless." In later years, this attitude even reached the universities, resulting in teachers losing "common sense of what kind of ignorance is unacceptable," as Lasch quotes a study of general education at Columbia. Such contentless education does not lead to the development of knowledge but to the development of opinion and the tools of opinion formation.

At the same time, family life, once a stern, true-north compass for values and behavior, diminished in influence on the individual. The loss of old certainties in work and social relations engendered doubt as to how to bring up children. Parents found themselves inhibited, incapable of instilling internalized disciplines and sharply detailed images of self and society into their children. Riddled with anxiety, parents turned to their contemporaries for advice, and each parent, while seeking and giving advice, found a source of direction in mass media, where society was communicating. Riesman concludes, "For in their uneasiness as to how to bring up children, they turn increasingly to books, magazines, government pamphlets, and radio programs"—which before television were the mass media.

It can't be emphasized enough that, in an other-directed society, personal pursuits and values are, in Riesman's words, "replaced by the fluctuating tastes" of the peer group that stands midway between the individual and the messages flowing from the mass media. What one ends up with, then, is everyone looking at everyone else who is looking at the media, which in turn is looking at everyone! The mass media become part of the peer group, stand-ins for the guidance and approval we expect from our contemporaries. It is a hall of mirrors. As Riesman puts it, "The mass media are the wholesalers; the peer-groups, the retailers of the communications industry. But the flow is not all one way. Not only do the peers decide, to a large extent, which tastes, skills, and words, appearing for the first time within their circle, shall be given approval, but they also select some for wider publicity through contiguous groups and eventually back to the mass media for still wider distribution." In an other-directed society more than any other, the media are simply and dramatically the vehicle of society communicating.

The social character of America when broadcast television became available was other-directed. Individual values and the style and norms of personal behavior were increasingly derived from sources outside ourselves. The agents of character formation, especially school and family, which had previously resisted the opinions and changing fashions of the external world, now became instruments of sensitivity training, the fundamental inventors and manufacturers of our internal radar systems. If Riesman and others are correct, all of this was true before television exploded on the scene.

Television: The Technology of Other-Direction

Born into a world of other-direction, television provided an ideal technology for the elaboration of other-direction in our daily lives. The question confronting the other-directed was ultimately epistemological: How can I as a member of society know what is relevant, meaningful, and useful? Individuals needed a common, sensitive field for their radar scanning. In this context, television became not only the critical vehicle of society communicating, but the key medium of society knowing.

Technology is an issue in the development of social character because it dictates how rapidly, how extensively, and how imitatably society communicates in the process of forming and affirming its personality. In these terms, television meshed perfectly with an other-directed society. Moving with electronic speed, television is instantaneous and simultaneous whether carrying the Super Bowl or a Senate hearing. Television is democratic, egalitarian, and pervasive because it is broadcast for all the public to see and hear. It is thoroughly mimetic because, like film and theater, it communicates the look, sound, and motion of physical life in an almost direct imitation of reality. For this reason, television reality can become part of our experience without the filter of self-reflection and without the need to identify and organize it as new experience.

The arrival of this technology was quite different from the arrival of printing. When the printing press created an audience, when the technology created a market, and general literacy developed, it helped support the inner-directed imposition of an ethic that served as a means of social cohesion. It provided a guide to experience and action. An other-directed consumer society could not and did not perform these functions when television technology suddenly became pervasive and radicalized every prior capacity of communications technology.

Television was far more pervasive and radicalizing than printing had been. It was massive. When Riesman and others spoke of books, magazines, and radio as mass media, they could not imagine the size and shape of television. There never had been a medium that could reach *everybody,* and reach them with images of behavior as behavior without the rationalization of words. The audience for its programs was drawn from every social class and every social element. By the mere act of watching television, a heterogeneous society could engage in a purely homogeneous activity.

Television images are more rapid and transient than the printed word. They make no demand on us to remember or reflect on them. This impermanence and the time of consumption cause us to spend extended hours with the medium but significantly less time with any one image or sequence of images. Television is instantaneous and simultaneous: Everyone gets the message at the same time and, *at the same time that an event is happening.* There is no lag time between a reporter witnessing an event and reporting it, and no time for reflection and analysis.

Because television can be universal, it must and will try to become so, both feeding and feeding on prior audience attitudes and beliefs. Its programming is significantly more open to vicarious experience than is print. Thus, if the vicariousness of print can powerfully produce the models that tell people what they ought to be like, as Riesman argued, then television has the inescapable capacity to be even more powerful for everyone all at once.

Inherently pervasive, inherently instantaneous, and inherently a mimetic image of behavior, television was perfectly suited for an other-directed society and became its stream of consciousness. Other-direction preceded television, but television technology propelled it and accelerated its social character, for the technology could offer what the audience demanded: easily digested and rapidly changing images of behavior. No other technology could deliver that as efficiently or effectively as television.

Technology is the issue, not program content, which is neither homogeneous nor capable of producing a homeostatic culture. The broadcast schedule includes baseball and ballet, symphonic and pop music, comedy and drama, car chases and inner crises, soap operas and major motion pictures, cartoons and news, in an ever-shifting kaleidoscope of sense and nonsense. It does not provide what everyone wants at every moment, but it has its moments for everyone. Television programming is fundamentally a mud-against-the-wall business, as writers and producers keep generating concepts and products until some stick with the public, even though most do not. Broadcast television as a business and a technology deals by its nature in the pervasive proliferation of images competitively acceptable to large and overlapping segments of society. In its use, broadcast technology has an epistemological effect. Much like Chauncy Gardener in the movie *Being There*, postwar America demanded and used the technology for base experience. In the process, the technology energized and accelerated other-direction in a way that could not have happened without it.

Riesman accurately predicted the other-directed personality's craving for and dependence on media, and television was not the only medium it devoured. Between 1951 and 1981 magazine circulation grew almost twice as fast as the population (92 percent versus 55 percent). Ninety percent of the adult population are magazine readers, and on the average they read almost eight magazines every month. On a per capita basis, twice as many books were sold in 1982 as in 1950. The typical adult today listens to the radio twenty hours each week, and 75 percent read a daily newspaper.

It was television technology, however, that was capable of becoming the primary, universal, and ubiquitous substitute for experience. It even created celebrities to be exploited by other media, as seen most dramatically in *People* magazine. Television technology also provided the forum for a staggering increase and inundation of images, generated by the constant, consistent flow of programming through four to seven available channels in each signal area, with every minute of every program filled with multiple images. Precisely because of its technology, broadcast television could and did become the central nervous system of other-directed society. It created, sent, and relayed behavior messages constantly, day after day.

Broadcast television technology was initially pure and indifferent. What has become of the technology is a result of our failure to imagine it and to build a social contract with it. What has become of us as a society is a result of how we, the members of that society, utilized the new technology. We were other-directed. We looked to television seriously for models of behavior and for opinions to fill our fundamentally opinionated environment. Television became the quintessential medium for a life-style of other-direction wherein our attention is on the ever-shifting attitudes, values, and judgments of our peers and wherein everything is news, momentary and fleeting, yet demanding action.

A Formula for Projecting the Future

The key question before us in analyzing communications technology in a given society is what difference each makes to the other. On the one hand, technology only represents possibility and potential. On the other hand, society is ultimately responsible for what it does with the technology. Furthermore, technology defines what *can* be done—and most probably *will* be done—by a particular society with a particular social character.

A similar case of technology and the social use of technology is the debate over handgun control. As a technology, the handgun is morally indifferent, as for that matter, is the act of using it. To this extent, the National Rifle Association's bumper stickers—"Guns don't kill people, people do"—are accurate. But it is precisely because of our thorough understanding of the potential of the technology, and precisely because of our understanding of human nature, that many argue for the control of the sale and possession of firearms. We have learned, the argument goes, what too many people will inevitably do with the technology that they couldn't do without it. The debate revolves around well-grounded understandings of both social character and technology, and what seems inevitable when they coincide. Our legislative actions are based on the social or moral value we put upon the inevitable results, which are really not up for debate.

The interaction of social character and communications technology also has predictable results. At work and play, parenting and consuming, other-directed people looked to others for both meaning and value. By this very act, they both dictated the content of communications and the meaning they received. If one accepts this description of our social character prior to the rapid and sudden pervasiveness of broadcast television, and if one can both imagine and accept the inherent characteristics of television technology—its contentless goals and techniques—one can and could have imagined the most probable future when the new technology impacted upon our social character and when we, as other-directed people, welcomed television and made it our primary way of knowing not only about the world, but about behavior.

As we reflect on the conjunction of other-directed social character and broadcast television technology, a working hypothesis for projecting our potential future suggests itself. This thesis is: *If you can understand the culture and social character of a particular time and place, and if you can understand the intrinsic characteristics of new communications technology, you can reasonably predict what will happen when technology and its users/consumers come together in substantial numbers.* Basically you can describe how they will energize one another, how the media will be used, and what the emergent social character might be like. Most important, the history of television suggests that the candid and sensitive application of this formula can provide an experiential understanding of the future capable of filling the lost imaginative space we have been confronted with to date.

The next step in applying this working hypothesis is an even closer, more detailed understanding of our current social character—the character that emerged from our other-directed use of television in its first explosive years.

3

Generating Today's Generation

Every generation holds within it the seeds of the next. The social revolution that exploded in the late 1960s and early '70s was the inevitable result of the American social character of the '50s. The seeds of that rebellion and the technology that energized and accelerated it were visible long before the rebellion occurred. The American social character of the 1950s, rooted in the conformism of other-direction, served by an extraordinarily healthy and vital economy, and stimulated by the televised cornucopia of factual and fictional experiences, bypassed any chance for gradual evolution and development. With a pace and intensity that took most of us by surprise, American society erupted suddenly and violently into the Autonomy Generation. The Autonomy Generation—often called the Me Generation—has laid the seedbed for the next era of social change—the Confetti Generation.

As we have seen, reality appears to other-directed individuals even today as a network of changing social relationships, and one's personal character is found in a corresponding series of roles. To this performing self, the only reality is the identity constructed out of communicated images. Television propelled this hyperkinetic search. The necessary and predictable result of this ungrounded inundation of images was a constantly shifting identity accompanied by a loss of self-value. Everyone was on his own because the images themselves had no grounding in time, place, logic, history, or knowledge. Being dependent on inner radar under such circumstances is like being constantly under surveillance. One tends to see life from the outside rather than from within, and the suspicion grows that one's inner core is a kind of illusion, a false secret kept from the world.

This state of anxiety led inevitably and predictably to the mass rebelliousness of the '60s and '70s and to today's self-absorption. Without the acceleration, multiplicity, fickleness, and pervasiveness of television's highly mimetic images, it is doubtful whether our other-directed social character would have so rapidly turned into the self-concern and self-assertion on which that rebelliousness was grounded. No other communications technology could provide such a repetitive and variable set of intensely stimulating images of behavior. No other medium possessed such potential for individual reinforcement derived from vicarious participation. Nor could any other medium so rapidly translate individual experience into massive sociopolitical action.

It is difficult to imagine the exact pace with which other-directed anxieties would have been radicalized if television had not existed. In fact, because the anxieties of other-direction were predictable, and because the dynamic of the technology was predictable, it would have been possible to predict what would happen if and when they came together, which they suddenly did in the short decade of the 1950s.

The core of the '60s revolution was an intense and anxious self-awareness. In the 1950s, vastly increased choices and social opportunities had created increased self-awareness, while other-direction failed to provide any sure or certain guide to personal behavior. The radar of other-direction had shifted each person's existential center from "society in general" to "other people like me," and shifted each person's source of knowledge from the clear certitudes of family, church, and school to a fickle peer group as reflected through the mass media.

Without clear certitudes to guide them and provide security, the other-directed began to feel a generalized anxiety. Rollo May argues in *Love and Will,* "In the late 1930's and early 1940's some therapists, including myself, were impressed by the fact that in many of our patients anxiety was appearing *not merely as a symptom of repression or pathology, but as a generalized character state.*" The world had become so filled with transient and contentless messages and images that personal communication based on honesty and truthfulness was exceedingly difficult and rare. In the other-directed environment, the proliferation and bombardment of images of behavior resulting from unrestrained technology were both the cause and condition of self-absorption, anxiety, self-assertion, and the loss of self-value. As May explained it, an individual frozen by the act of choice, too insecure to take a chance, "is forced to turn inward; he becomes obsessed with the new form of the problem of identity, namely, Even-if-I-know-who-I-am, I-have-no-significance." In other words, anxiety had become a way of life for a significant number of American citizens.

John Kenneth Galbraith has argued in *The Affluent Society* that "if the individual's wants are to be urgent, they must be original with himself. They cannot be urgent if they must be contrived for him. And above all they must not be contrived by the process of production by which they are satisfied." He was thinking in economic terms, but the same formula applies intellectually in an other-directed society where the mass media create not only consumer but behavioral needs and wants and in turn promise to show us how to fulfill them. At some point in such a pressurized situation, the volcano will erupt. The revolution in social character that emerged from the other-direction of the '50s combined intense self-awareness with the insistence that one's wants and needs be original with oneself and entirely uncontrived. Or as the popular slogan of the '60s put it, "Do your own thing."

We have come to accept this self-aware self-assertion as part of the definition of our age. In fact, the Me Generation that has grown up since World War II has arrived in three waves: a vociferous self-assertion characteristic of the '60s, which led to an intense self-absorption in the '70s, which finally settled into an abiding sense of autonomy in the '80s. Yet these shifts in social character would not have happened as rapidly or intensely without the pervasive bombardment of images of behavior made possible by television technology. Both Riesman and May predicted that as we were forced to market ourselves at home, at school, and at work, we would eventually become acutely aware of the self as being without anchor in an other-directed environment. Adrift in a sea of anxiety and eventually recognizing the fickleness of other-directed values, we would naturally devalue values themselves and assert our individual selves as a way of escaping from that anxiety.

It comes as no surprise, then, to read in Yankelovich's report that "in the nineteen seventies all national surveys showed an increase in preoccupation with the self." It was inevitable. "By the late seventies," Yankelovich continued, "my firm's studies showed more than seven out of ten Americans (74%) spending a great deal of time thinking about themselves and their inner lives." An other-directed person who becomes aware of self precisely because of the constant effort of other-direction and who is faced on all sides by constantly changing images of behavior is almost bound to end up in a ceaseless, anxious search for the real self. At some point such a person will announce that presumed "real self" to the world. Communications technology had simply made the burden of self-identity too great. Television, uncontrolled at its personal reception point, and uncontrolled by larger social forces, made the difference.

The Troubadours of the Rebellion

In his introductory essay to *Existential Psychology,* Rollo May cites William Whyte's concern that the enemy of modern man may turn out to be a mild-looking therapist who induces conformity while thinking he is helping us. Whyte refers, he says, "to the tendency to use the social sciences in support of the social ethic of one's historical period." In fact, something like this occurred with the existential psychologists, especially with Carl Rogers and Abraham Maslow—only it occurred from the bottom up, rather than from the top down, as both men pieced together their analysis of the American social character directly from encounters with their clients.

Alexis de Tocqueville wrote that "in no country in the civilized world is less attention paid to philosophy than in the United States." That may have been true in the 1830s, but in absorbing and identifying with the works of Rogers and Maslow, a critical mass of modern Americans internalized them as their philosophers. They did so not in the sense of ultimate truth or unified dogma, which most Americans believe is unattainable, but in the American pragmatic sense of discovering a partial truth that worked for them. In this context, Rogers and Maslow became the troubadours of our current social character, even if they offered, in Yankelovich's words, a "falsely seductive" appeal and "a peculiarly self-congratulatory philosophy for a materialistic age."

The existential psychologists based their conclusions on therapeutic experience with clients and then found a sympathetic response in a wide range of people who identified with the experiences and feelings of those clients. Generally, they discovered and focused on the prevailing anxiety in a rapidly changing society over the loss of those goals and values that individuals had previously relied on to determine self-value. The loss of accustomed values, coupled with a vacillating response to confused and varying social signals radicalized by communications technology, produced a loss of self-dignity and identity. As a group, these therapists encouraged an active rather than a reactive approach toward growth. They emphasized a view of the person as an active, creative, experiencing human being capable of allaying anxiety and rendering experience meaningful. To them Freud was the high priest of inner-directed guilt, whereas they were absolving not guilt, but the anxieties of the other-directed.

The existential psychologists drew intellectual sustenance from Carl Jung's emphasis on human nature's positive tendency toward growth. Like Husserl they were absorbed by the study of the "given" of immediate experience. Like Martin Buber, they came to the conclusion that understanding through encounter is as real as, or more real than, understanding through objectification. In counseling, they increasingly emphasized that what is important is not the object or event in itself, but how it is perceived or understood by the individual. Thus they stressed both immediate experience and subjectivity as the primary modes of awareness and knowledge.

Their values became honesty, genuineness, and authenticity. They affirmed, in the words of Søren Kierkegaard, the need "to be that self which one truly is." They encouraged their followers to determine reality for themselves, to be total individuals and, by so doing, establish their own moral imperatives. Thus their recommendations ended up being more like Sartre's insistence on carving out one's own existence quite simply by what one does.

Using the preceding overview as a starting point, let's look at these heralds of our current social character individually and discover how their own words tapped the anxieties of a generation of other-directed Americans and gave them hope and expression.

Carl Rogers was perhaps the most paternally encouraging of the popular psychology stars of the time. His simple conversational writing style and his conscious development of "warmth and emphatic understanding" were easy for the lay person to understand. Rogers's message reached far, from the most to the least anxious, and consequently reflected the widest spectrum of current social beliefs. In general, his therapy encouraged people to move away from façades, away from "oughts," away from meeting cultural and social expectations, and away from pleasing others. He gently guided his listeners toward greater self-direction, perceived as an ongoing process of opening oneself, of experiencing life, of accepting others and trusting oneself.

In the popular mind, these steps became the stages of growth. As he states in *On Becoming a Person,* "First of all, the client moves toward being autonomous. . . . Less and less does he look to others for approval or disapproval; for standards to live by; for decisions and choices. He recognizes that it rests within himself to choose; that the only question that matters is: Am I living in a way which is deeply satisfying to me, and which truly expresses me?" Thus the client "becomes responsible to himself. He decides what activities and ways of behaving have meaning for him, and what do not."

The person who emerges from Rogers's therapy has lost defensiveness and is open to personal experience. He "increasingly discovers that his own organism is trustworthy, that it is a suitable instrument for discovering the most satisfying behavior in each immediate situation." Consequently, he realizes that "what I will be in the next moment, and what I will do, grows out of that moment, and cannot be predicted in advance either by me or by others." There is an increasing openness to experience, and "he is free to live his feelings subjectively, as they exist in him." In short, the good life means living fully in each moment.

Rogers's concern was that "in choosing what course of action to take in any situation, many people rely upon guiding principles, upon a code of action laid down by some group or institution, upon the judgment of others (from wife and friends to Emily Post), or upon the way they have behaved in some similar past situation." He argued, on the contrary, that we should be open to experience and if we are open we will discover that "doing what 'feels right' proves to be a competent and trustworthy guide to behavior which is truly satisfying." We need not worry about who will socialize such a person, "for one of his own deepest needs is for affiliation and communication with others. As he becomes more fully himself, he will become more realistically socialized."

In summary Rogers approvingly quotes from a letter from a client who had successfully completed the process: "I've always felt that I *had* to do things because they were expected of me, or more importantly, to make people like me. The hell with it! I think from now on I'm going to just be me—rich or poor, good or bad, rational or irrational, logical or illogical, famous or infamous." Here is a cry of release from the anxieties of other-direction, an assertion of self that transcends any set of values. With "warmth and emphatic understanding," Rogers supported this process as therapy, as health, for to him the ultimate value—the good life—was to be oneself.

In contrast to Rogers's approachable, paternal warmth, Abraham Maslow exuded the challenge of an older, successful sibling, energizing his readers with a spirit of emulation. There was a sense of challenge in his assertions of hierarchy and the urge to self-actualization that many found impossible to resist.

Speaking directly to the anxieties of other-direction, Maslow pointed out in *Toward Becoming a Person* and *Motivation and Personality* that only other people, persons outside ourselves, could satisfy needs such as belongingness, respect, and love relations. The anxieties of other-direction certainly confirmed that! Other people were in control of our lives, not ourselves. As Maslow put it, "Their wishes, their whims, their rules and laws govern him and must be appeased lest he jeopardize his sources of supply. He *must* be, to an extent, 'other-directed' and *must* be sensitive to other people's approval, affection and good will . . . *he* must adapt and adjust . . . *he* is the dependent variable." Such a person must be "afraid" of his environment, and "we now know that this kind of anxious dependence breeds hostility as well."

For younger people experiencing the anxieties of other-direction during and after college, Maslow's positive prescriptions were existentially attractive and challenging. From the beginning, Maslow based his prescriptions on healthy people who had made it and he argued backward and downward to the source of the problem. He spoke as one liberated to the unliberated. "I reported (in a 1951 paper) my healthy subjects to be superficially accepting of conventions, but privately to be casual, perfunctory and detached about them. That is, they could take them or leave them. In practically all of them, I found rather calm, good-humored rejection of the stupidities and imperfections of the culture with greater or lesser effort at improving it." Maslow insisted that almost every serious description of the "authentic person" revealed an individual who assumed a new and unique relation to his or her society and to society in general. In Maslow's words, "He not only transcends himself in various ways; he also transcends his culture. He resists enculturation." Simply put, the healthy person extols a Byronic rejection of personal anxieties and of the life-style of other-direction.

His healthy clients had taught Maslow "to see as profoundly abnormal, or weak what I had always taken for granted as humanly normal; namely that too many people do not make up their own minds, but have their minds made up for them by salesmen, advertisers, parents, propagandists, tv, newspapers and so on." He also had little doubt that "historically, we are in a value interregnum in which all externally given value systems have proven to be failures—political, economical, religious." Against this background "the person, insofar as he *is* a real person, is his own main determinant." Such real persons are autonomous, "ruled by the laws of their own character rather than by the rules of society (insofar as these are different)." They are ethical if unconventional. "The unthinking observer might sometimes believe them to be unethical, since they can break down not only conventions but laws when the situation seems to demand." But the very opposite is the case. They are simply alienated from "ordinary conventions and from the ordinarily accepted hypocrisies, lies and inconsistencies of social life."

In principle what makes education, self-improvement, and psychotherapy possible is "the will to health, the urge to grow, the pressure to self-actualization, the quest for one's identity." The consequence of betraying one's own inner nature or self, or skipping away from self-actualization, is intrinsic guilt and self-disapproval. Maslow argued in the '50s and '60s that it was best to recognize, bring out, and encourage this inner nature, rather than to ignore or suppress it. "Pure spontaneity consists of free, uninhibited, uncontrolled, trusting, unpremeditated expression of the self." To be free, we must not fear our own psyche but express it and act upon it.

Maslow's message went directly to the heart of anxiety-ridden Americans living in relatively affluent times. He candidly admitted that "the better culture gratifies all basic human needs and permits self-actualization. The poorer cultures do not." The assumption in the United States at the time was that we had such a successful culture, and, in fact, it would have been difficult to argue the opposite with the children of suburbia.

As troubadours of the Autonomy Generation, both Carl Rogers and Abraham Maslow provide us with the voice, the language, the attitudes, the perspectives, and the reasons why we act out our contemporary social character. As one of the principals of the outstanding consumer-research firm Yankelovich, Skelly and White, Daniel Yankelovich monitored the growing resistance to enculturation that in our times has become a counterculture of its own. Yankelovich accumulated an enormous data bank on our habits and attitudes during the last three decades that enables us to check our assumptions, to both verify and quantify behavioral trends.

In *New Rules: Searching for Self-Fulfillment in a World Turned Upside Down,* Yankelovich reports that "in the 1960s the search for self-fulfillment was largely confined to young Americans on the nation's campuses," but that "the challenge to traditional mores spread beyond college life to find a variety of expressions in the larger society." Summarizing the data, he concludes: "By the seventies, most Americans were involved in projects to prove that life can be more than a grim economic chore. Americans from every walk of life were suddenly eager to give more meaning to their lives, find fuller self-expression, and add a touch of adventure and grace to their own lives and those of others. Where strict norms had prevailed in the fifties and into the early sixties, now all was pluralism and freedom of choice.''

It is not surprising that three out of four Americans today, still reflecting the legacy of the '60s, believe that they have more freedom of choice about how to live their lives than their parents did. What may come as a surprise is that "an overwhelming 80% feel confident that they will be able to carry out these choices" and live their lives any way they choose. Curiously, however, there is an equal portion of cynicism amid this confidence. "The majority of Americans (81%) have come to suspect that those who follow the rules inevitably get cheated while those who know the angles and ignore the rules do well." On the face of it, this is a striking combination of ends and means, but Yankelovich also reports a strong increase in recent years in aimlessness, especially among young people, and a growth in hedonism among people who abandon the old rules without adopting new ones they can believe in.

Yankelovich concludes that their

> exuberance is shot through with anxiety. Willfulness is coupled with a crippling sense of powerlessness. Decisiveness is undercut by confusion. And the warmth of self-acceptance is muted by a strange feeling of estrangement from the world. They are not sure how to choose and they are reluctant to risk freedom by making commitments that may prove irrevocable. They seek to preserve freedom by failing to risk it. They do not see themselves as part and parcel of an ongoing social world, progressively discovering themselves in relation to their work, their friends, families and the larger society. Rather, they are isolated—some might say existential—units, related intimately only to their own psyches.

Christopher Lasch has summarized the dilemma later quantified by Yankelovich, even describing a situational ethic for people caught in the culture of narcissism. He writes, "The narcissist has no interest in the future because, in part, he has so little interest in the past . . . to live for the moment is the prevailing passion—to live for yourself, not for your predecessors or posterity."

The thread of social character that we have been pursuing suggests that the seeds of self-concentration were rooted in other-direction, and that the pace, pervasiveness, and panic of the experience of other-direction were supported by television technology, with its inherent, kinetic multiplicity of images of behavior. It simply could not have happened in the same way without television. The existential psychologists gave a supportive, shaping voice to the predictable anxiety while framing the widely accepted solution of self-actualization.

Neither moral outrage nor chagrin are relevant to our concern about social character. What is important are its firm-rooted existence and characteristics so ably quantified by Yankelovich's data. He concludes, "The self-fulfillment search is a more complex, important, and irreversible cultural phenomenon than simply the by-product of affluence, or a shift in the national character toward narcissism." There seems to be no turning back. Our current social character is deeply rooted. We are who we are.

The Autonomy Generation

The evidence is overwhelming: Someplace in the 1950s through social mobility we lost the past, and certainly by 1970 through self-absorption we had lost the future too. We became individuals with no particular place in history, looking neither forward nor backward, with no particular obligation to either the dead or the unborn. The worldview rooted in our current culture and derived in great part from Abraham Maslow and Carl Rogers emphasizes the individual person as an active, experiencing human being who lives only in the present, responding to current perceptions, relationships, and encounters, a person who is insistent that understanding through encounters is more real than understanding through objectification. Immediate experience is the given, but what is important is not the object or event in itself, but how it is perceived and understood by the individual. Reality is essentially a very private matter determined by individual perception. It is taken for granted that the primary motivating force of humankind is the drive to self-actualize, and in its current narrow interpretation, this means that relationships are posited or denied, accepted or rejected, on the single basis of their perceived impact on one's autonomous goals.

Currently we are experiencing the Autonomy Generation, which assumes that it has nothing to learn from any prior human experience or consensus. We are reluctant to accept authority. We suffer fractionated identity crises. We are living in a generation that contradicts itself. On the one hand, we believe that our perceptions are inevitably wiser than those of both our forebears and our neighbors. On the other hand, in our current state of disillusionment we believe that human judgment is so idiosyncratic and institutions so irrational that it is impossible to reach any trustworthy objective judgments. The best we can hope for are merely personal conclusions about anything. We are also a generation anxious to simplify the sheer complexity of contemporary life. We are a generation armed with computer literacy and an openness to technology, a generation abused by the energy crisis and desiring the substitution of cheap communications for expensive transportation. In short, a generation quite different from that which confronted television technology for the first time and purchased the first fifty million television sets in the 1950s.

Today's Everyman doesn't have an internal gyroscope like the inner-directed American of the pre–1950s, nor a radar dish and screen like the other-directed Organization Man and his family of consumers. Rather the Autonomy Generation has a set of scales. On one side of its scales is who they assert they are today, this week, this season, this year. On the other side are weighed all their other experiences in an almost undifferentiated flow. Only those that temporarily balance are temporarily accepted.

It was not a war of worlds or of worldviews. It just happened. Once television became pervasive in a society dependent on radar scanning for its culture, it became definitive. It became a gestalt without form. It mesmerized a lonely crowd obsessed with accommodation and fragmented it into self-concerned, self-centered individuals. Such an outcome was predictable. With a prudent presence of mind, a knowledge of the relationship between communications and society, and a clear perception of the intrinsic characteristics of television, it would have been possible to predict the culture of the "Assertives" and the "Self-Directed" who became and are the Autonomy Generation.

If you know a particular culture's mind-set, and if you know a communication technology's inherent characteristics, you can predict what will happen when they come together. You can predict future cultures and markets as articulated in social character. As we learned with television, the cultural effects of a communications technology can be in evidence long before the media are clearly pervasive in society, and as we shall see very shortly this will be even truer of the new electronic media than of television. The new electronic media will be as congruent to our Autonomy Generation as television was to the generation that preceded us.

To apply the formula accurately, we need a clear understanding of each of the new technologies. We need to know what they are, where they come from, how they are understood, how they are being used, where each one appears to be going, and how far it can go. To make the imaginative leap to describe the impact of these technologies on our social character, we must stand on a bedrock of reality and facts. We must fill in the lost imaginative space with precise description, business history, government regulation, market research, artistic efforts, personalities, and even dreams. Only then can we confidently project what will happen when the Autonomy Generation gets these technologies in its hands.

PART 2

THE
TECHNOLOGY
CONNECTION

Twenty years ago when Marshall McLuhan announced that "the medium is the message," he argued that it is a "practical fact" that technology, not what we do with it, alters our culture, society, relations, and intercourse. The very technologies of communications cause social change, regardless of their content or individual use. Certainly one of the compelling facts of history is that major developments in communications technology create, or cause, new social structures to come into being. This was true of writing, printing, and broadcasting, and it will be true of the new electronics.

What follows from this "practical fact" is the necessity of understanding these technologies in and of themselves. If a communications technology has a predictable impact on society regardless of what we do with it, then it would follow that our effort in the observation of current phenomena should be to discern the intrinsic characteristics and, therefore, the predictable effects of these technologies.

Understanding any medium or communications technology

is difficult enough, but the complexity and multiplicity of to-day's media compound the problem. Television, by these stan-dards, was simple. The key fact is that all of the new electronic media—satellites, broadcasting, cable and pay television, video-cassettes, videodiscs, personal computers, and interactive vid-eotex systems—and all the communications products and services they offer will be simultaneously available to the American citizen consumer in his home. All of these media are the message, and they will come upon us with the same sudden-ness and revolutionary effects as television. In an effort to grasp this message, we will examine each of the new media in detail with as many historical, business, and creative facts as possible.

To achieve a realistic understanding of the future whether personal or social, economic or cultural, we must understand the intrinsic characteristics of the available communications me-dia. As we should have learned from broadcast television, the intrinsic defining characteristics of media dictate what they can and will be, what they provide and do, and how, in effect, they will insist on being used. It is possible today to know these things about all of the new electronic media not only in and of themselves, but especially in the ways they will coexist compet-itively. A medium, like any other tool or business, will not survive attempting to do what another does better or more economically.

In general, communications media can be organized as an inverted pyramid. Media designed to serve nearly everybody at the same time, such as television, are at the top, and those designed to serve individuals one at a time, such as the tele-phone, are at the bottom. Between the broadscale media at the top and narrowly focused media at the bottom, all media can be assigned a place in a competitive environment.

In the past we have occasionally been misled, because in the absence of competition, media will often perform functions in society for which they are not ideally suited. It is only after a more appropriately designed technology arrives and achieves sufficient distribution that we are alerted to what was inevitable at the time of invention. This is precisely what happened when mass magazines and television began to compete. The outcome was both inevitable and knowable, but many of us were easily misled at the time because of the way some magazines had functioned in the absence of competition and because we failed to imagine television's widespread success. Much the same story can be read in the history of painting and photography. Hopefully by applying what we should have learned from the past in a sane yet imaginative way, we can develop a realistic picture of the new electronic media.

4

Cable and Broad-scale Media

For a long time what we have come to call cable television was a quiet, small-town business. About six months after L. E. Parsons first strung his television antenna wires over the mountains outside of Seattle in 1949 so his wife could receive network programs, the Associated Press discovered his initiative and *The Washington Star* printed the story. Staffers at the Federal Communications Commission read the story, sent Parsons an inquiry, and, apparently satisfied with his reply, approved his business. Interestingly, from the very beginning all of the active players were present: a neighborhood that wanted the television that everyone else in the nation received, an entrepreneur who invested in receiving equipment and wires to provide it, and, of course, a government anxious to assert its prerogatives and regulate what it called an "extension" of broadcasting. Yet for the first fifteen years this was strictly village trade.

It was a small-town business. In 1950 Robert Tarlton joined with Milton Shapp (who later became governor of Pennsylvania) to build what was the nation's second cable system in Lansford, Pennsylvania. It was here in the hills of Pennsylvania that cable television was invented as a community medium and achieved its first significant penetration. And not long after that, Martin Malarkey founded a cable system twenty miles southwest of Lansford to help sell television sets through his father's music and appliance store—much as David Sarnoff founded broadcasting to sell radio sets. In 1952 these and other entrepreneurs founded the National Community Antenna Television Association, representing nineteen systems. They derived their name from the phrase "Community Antenna Television," which Al Warren, the editor of *Television Digest,* coined because "it was indeed a hamlet service, usually the sole source of TV in town." The business grew in precisely this way until in 1965 there were 1,570 systems providing service to 1,575,000 subscribers, or 3 percent of U.S. households.

No one took much notice of cable television in those days, not because of the modest number of subscribers, but because of what it was and how it functioned. We are accustomed to believing that broadcast television is free, but it isn't. Not only is advertising support of programming passed along to the consumer as part of the price of goods and services, but, more important, access to the medium is a relatively costly investment. The average television set costs $500, and many people in the 1960s paid considerably more than that. The purchase and maintenance of a television set is similar to the purchase of a ticket to the theater or a ball game, or to the purchase of a magazine subscription. It is the cost of admission, the cost of access. Most people paid for this access at the rate of a few dollars per month—in time payments for the television sets and in modest increases in their electric bills.

People living in remote or rural areas, accustomed to paying a few extra dollars for most products transported over the hills to their towns, paid a few extra dollars per month to receive television programming via cable. Broadcasters and regulators were pleased because cable created local communications equality and increased network and station audiences by the national or local cable penetration. Cable television was simply an alternate distribution system for the same products in the same markets, using a technology not much different from your home antenna wire.

Several lessons can be learned about the nature of cable television from this brief history. First, cable is a distribution system based on the same products as broadcast television, which is also a distribution system. Second, consumers everywhere have always invested in the right of access to this product. Third, both broadcasters and regulators perceived cable distribution as an extension of broadcasting.

Cable television grew because Americans were by no means equal in communications terms, nor did their purchase of a television set bring an equal return on investment. People in New York and Los Angeles, for example, had five times the number of broadcast channels as most Americans. In fact, because the three broadcast networks and their affiliates were without competition in many areas, they generally enjoyed over 95 percent of the viewing.

New cable technology could and did make Americans equal. It could distribute multiple signals—at first three or five channels, and ultimately twelve or more with the newly developed coaxial cable. In addition, it could provide product inexpensively by importing broadcast signals through a second new technology—microwave transmission—from other markets where they existed. To reflect these technological and service developments, the National Community Antenna Television Association changed its name to the National Cable Television Association. A major legal debate developed as broadcasters fought to protect their market by arguing that cable systems had no legal right to move broadcast signals from one market to another. In larger markets where independent stations existed, the network and affiliated stations' share of the audience was 10 percent to 20 percent less than where they held a monopoly. Furthermore, the independent stations involved saw no value in the distant or extended audience made possible by cable distri-

bution because their revenues were derived from local marketers such as automobile dealers and banks, or from national
advertisers who defined their markets by broadcast signal areas—neither of which groups would pay extra for audiences
outside of these designated areas.

Generally representing broadcaster interests, the FCC in 1966
imposed a freeze on the importation of distant signals via cable.
In 1972, after six years of acrimonious debate, the commission
lifted these restrictions and encouraged cable development. The
freeze thawed because cable had found a constituency among
the public in congressional districts, a constituency founded on
the issue of equality of citizen television service and on consumer willingness to pay a few extra dollars per month to improve the access already invested in by the purchase of a television set. Cable proved to be the most efficient way of
developing an abundant multichannel television service to the
American consumer. As 1972 came to an end, there were 2,991
cable systems with 7.3 million subscribers, or about 11 percent
of American households. In seven years, cable had doubled the
number of systems while quintupling the number of subscribers
and was on the verge of becoming a billion-dollar industry.

The debate, in the meantime, generated many of the fantasies
about cable television that continue to distort our vision of the
future. The broadcasters and their supporting industries, trumpeting unfounded fear, had based much of their argument on
the public interest expressed as the "pity curve." They made the
case that news and public-affairs programming, as well as occasional quality children's programs, did not pay their way and
were sustained in the public interest by profits diverted from
entertainment programming. If these profits were diminished
by siphoning off their audience to fragmented cable programming, the result would be a weakening of news, documentaries, and quality children's programming. Obviously, they argued, this would not be in the public interest.

The opponents to these claims, trumpeting unfounded hopes,
projected cable as the medium most in the public interest

because of its ability to satisfy specialized interests of every kind through narrowcasting, public-access channels, local origination of community programming, and two-way communications for everything from burglar and fire alarms to data transmission. In the process, cable television became the hope chest of everyone's expectations for every kind of electronic communication. It was all a fantasy, bred in the heat, rather than the light, of debate. These fantasies bedeviled our conceptualization of the medium, fragmented our imagination of the business, tormented the franchise process and regulatory discussions, and even led the broadcast networks to invest and lose millions of dollars in quixotic cable enterprises.

On November 8, 1972, Home Box Office turned one switch in New York, and John Watson, at the other end of a microwave link, at the head end of his cable system in Wilkes-Barre, Pennsylvania, turned another switch, and 365 cable television subscribers watched a Paul Newman–Henry Fonda movie, *Sometimes a Great Notion.* For a fee. HBO offered a scrambled signal of first-run movies without commercials. It was not a basic service that automatically came with a cable subscription. A subscriber had to pay extra to have the signal unscrambled. HBO was the first successful pay television service.

In some respects, however, HBO was born not with the turn of a switch in 1972, but with the launch of a rocket from Cape Kennedy in 1975. HBO was the first programming service to utilize a communications satellite for network distribution— though it would not be alone for very long. It was the use of satellites that finally interconnected thousands of cable systems and made cable television an organic whole. Like the second son's Easter suit, satellites were not new, but they were new to cable.

Two elements made satellites the perfect partner for cable systems. Satellites are "distance insensitive," meaning that, unlike land lines, the cost of using satellites was the same no matter whom you talked to, or how many times, or where. In this respect satellite use resembled broadcasting, where transmission cost the same no matter how many television sets filled the community. The cost of satellite transmission does not increase with distance or with the number of cable systems receiving it. Furthermore, satellites have great capacity—twenty-four channels or transponders each. This capacity appeared like a new frontier, and an Oklahoma land rush began. The base cost of distribution had come within the $1 million range—well within the reach of many American entrepreneurs. All they had to do was raise a hand and stake a claim.

Home Box Office staked the first claim in 1975, and a year later it had leaped from 57,000 subscribers in four states to over 282,000 across the country, or fifteen times its previous annual growth. Unlike previous pay-television attempts, HBO was not pay-per-view, nor did it function on a leased channel. It was built on a monthly consumer subscription fee, and it shared 50 percent to 60 percent of its revenue with the system operator. Hollywood's share as the supplier of the movies that formed the basis of HBO's programming would fluctuate between 15 percent and 30 percent and cause a great deal of contention. But HBO operated profitably at 10 percent to 20 percent of the gross for its middleman service. HBO was joined on the satellite in 1978 by Showtime and in 1979 by the Movie Channel, each with a slightly different set of programming strategies. These services have since merged, and other pay services, from the Playboy Channel to the Disney Channel, have emerged. But ever since cable systems were joined via satellites with HBO, 40 percent of all the consumers who can identify cable television, identify it with Home Box Office. Thus, in the consumer mind, cable is identified with movies, the most popular broad-scale entertainment product in our history.

Video Publishing

In the control room on September 27, 1977, it looked like any other Rangers–Fliers preseason hockey game at Madison Square

Garden. The commercial for BMW automobiles was counted down and inserted at the first break, just as on broadcast television. Then the phone rang. As in the early days of radio, it was Pensacola calling: Their signal was coming in loud and clear on the Florida cable system. In this manner, the first advertising-supported cable satellite network was born.

The network was a joint venture of UA-Columbia Cablevision and Madison Square Garden, supported by the efforts of the advertising agency Young & Rubicam. Beyond its use of the satellite to show the way for others, beyond the modesty of its beginnings, and beyond the striking fact of how few years have intervened between now and then, the Rangers-Fliers game was a watershed. The modest venture, called "video publishing" by the partners, was truly revolutionary. Up to that time whatever the service being employed—television, telephone, radio, or the Postal Service—the sender paid for the right of access. For example, the twenty-odd cents a consumer pays for a stamp or a telephone call is a payment for the right of access to the distribution system. In television and radio, the networks pay their affiliated stations a negotiated percentage of their national advertising revenue to insure that the local affiliates carry their programming. The pay television networks operate in effectively the same way with their cable system affiliates who pocket half of the subscriber revenue.

For the first time, the sender, in this case the Madison Square Garden Network, *required* a fee from its cable affiliates, the middleman distributors, at the same time that the network was also receiving revenue from national advertisers. In other words, the network had two revenue streams—from the subscribers through the cable system, and from advertisers. Publishers of magazines and newspapers also have two revenue streams—from the consumer/subscriber and from advertisers—which is why the partners called it video publishing. It was this act of charging their affiliates that defined cable television as different in kind from all previous electronic communications.

After a complicated history strewn with bankruptcies and red ink, all of the successful cable networks now charge a subscriber fee of five to twenty-five cents per subscriber per month for their advertiser-supported networks. One clear reason is economics, for at this time advertiser revenue does not cover the cost of programming and satellite distribution. But it is this revolutionary change in payment relationships that allows cable television to be different from broadcast television.

As a concept, video publishing goes beyond the strictly economic without excluding it. It is in the cable operator's interest to pay a fee for his programming for it gives him control over his channel capacity and the ability to shape his programming mix to the needs and expectations of his subscribers. For the cable operator, programming is marketing and affects the number of subscribers he signs up and the prices he can charge. If he did not pay for his programming, the cable operator would be forced to become a common carrier like the telephone and the satellite, forced to provide his channels to whoever wishes to pay the government-dictated rate. He would lose control over his own business and the rate of return on his investments.

Video publishing is also in the consumer's interest; for, as with magazines, a subscription insures the producer, editor, and publisher against servitude to advertiser interests. It is the consumer's subscription that enables *Time, Newsweek,* and *U.S. News & World Report; Forbes, Business Week,* and *Fortune; Good Housekeeping, Woman's Day,* and *Redbook; The New York Times* and the *Daily News* to be different from one another. Advertising is intrinsic to their viability as rather broad-based print vehicles, but it is the consumer's subscription that supports their independence, their variety, their distinctive shapes, and their editorial voices.

Video publishing is, in addition, in the communicator's interest as well, not only because it makes communications possible, but also because there is not a communicator worthy of the name who does not wish to be free from the demands of homogeneity.

Finally, video publishing is in the advertiser's interest because it provides communications product more precisely attuned to various consumer interests than any comparable vehicle totally supported by advertising, and consequently a vehicle more precisely attuned to targeted markets.

None of these are virtues of broadcasting, nor, as we have seen, could they ever be. But through the joining of video publishing with cable distribution, consumers can enjoy single-subject vertical channels such as ESPN's (Entertainment and Sports Programming Network) twenty-four hours of sports or CNN's wall-to-wall news.

The Narrowcasting Illusion

One might imagine that when the concept of video publishing is applied to television networks the result will be the narrowcasting of special-interest programs, such as "Runner's World" or "Beehive Management," on hundreds of cable channels. Nothing could be further from the truth. In 1971 the Sloane Commission on Cable Television suggested in a report that the analogy for cable television "is not to conventional television, but to the printing press." This commission made an important and valuable contribution to the debate over cable regulation at the time, but in seeming to draw a close connection to the printing press, it supported a misconception of cable. As the commission itself warned, "the analogy cannot be pushed too far; like all analogies it has its limits."

The first difference cited was the cost of production. "A few dollars at a job printing shop will buy a thousand flyers; there is very little in television that a few dollars will buy," noted the commission. Beyond the lights, cameras, and other equipment, television is at root a theatrical production requiring the collaborative work of a great many people over extensive periods of time. A book can be printed and published for less money than the most stripped-down half hour of television costs. Consequently, there is an intrinsic, inevitable "cut-off point for cable television that must reduce, in some degree, its flexibility and its real copiousness"—as distinct from what it might possess in theory, "and dramatically distinct from what print can provide."

Television of any variety is transient, hence different from printed matter. It does not admit of boredom, for it cannot be skimmed like a boring book or magazine. Not only is bad television, like bad theater, intolerable, but it cannot be set aside for consumption a chapter at a time if the material is too demanding. There is, according to the Sloane Commission's report, "a profound difference between a product such as a book that persists over time and one that is as evanescent as a television presentation. The television production makes a single appeal to the entire potential audience, and in immediate competition with other appeals." Because cable television in conjunction with the satellite reduces the cost of distribution, because of channel capacities, and because of direct subscriber support, cable television can be one or two steps removed from the demands of the mass audience that rules in broadcasting, but cable is still an "everybody medium."

Cable television creates a competitive environment for its products that forces the subscriber to draw the line at what he is willing to pay for. To make the economics of this evanescent and transient medium work, its programming and audience must be conceived of broadly, in order to provide a fairly large number of subscribers with simultaneous satisfaction and community of interest. Cable television has many characteristics in common with broadcasting, and that is why, with the possible exception of a town where half the population are beekeepers, nothing remotely resembling "Beehive Management" or any such programming will ever find a home on cable television. If narrowcasting implies the distribution of programming of possible interest to only decimal-point percentages of the viewing population, cable television is not the appropriate distribution system. On the other hand, if narrowcasting implies that cable programming can function successfully at only 20 percent of the interest levels required by broadcasting, then it is a useful word, appropriate to the technology. Unfortunately, as most commonly used, narrowcasting is a chimera.

Predicting the Growth of Cable Television

For decades, the prediction of cable-television numbers has been like attempting to predict politics, the economy, and the weather all at the same time. Politics, because cable is a regulated business. The economy, because cable is both a capital-intensive business and a consumer retail business. The weather, because cable is a construction business. Yet, based on a hard-nosed look at the facts, it has been possible to make remarkably accurate predictions of subscriber growth, as well as of individual cable network growth and performance.

The success of these prognostications has not come from mystical powers. Instead, unencumbered by fantasy, successful projections result from a close examination of all the relevant data, from franchising to construction investment to marketing performance to program audiences and, most especially, from a hands-on understanding of the medium and its intrinsic characteristics. It is possible to know, for example, how many cable subscribers there are today, when their systems were built and how many of their neighbors in the franchise area have chosen not to subscribe to cable. It is possible to know how long it usually takes to negotiate a franchise through a city council, and how many cities have begun the process, have completed the process, or have not begun at all. It is possible to know how long it takes to construct a cable system of a specific number of miles under specific rural or urban conditions. It is possible to know the channel capacity of all existing systems and of those where franchises have been granted and construction has begun. It is possible to know the political climate of franchise negotiations and, therefore, to have an informed judgment of the probable outcome in channel capacity for new systems and the re-building of old systems. It is possible to know the history and to measure the present success of marketing efforts, and, there-fore, to know approximately how many of the homes passed by the cable will become subscribers, and at what price and for what services. It is possible to know what consumers watch on television and on cable, and what they feel that they are miss-ing. In short, it is possible to ground our imaginations of the future in a solid foundation of fact.

It is not my purpose here to detail and enumerate each of these factors on complicated charts and graphs. What is rele-vant, however, is the recognition of how knowable these fac-tors are, how intransigent the knowledge of them is, and how little they admit to speculation. Those who seriously study all these knowable factors about cable television evince a great deal of consensus and very little dispute on the general shape of cable over the next five to ten years.

The consensus is that in 1990 there will be approximately 95 million television households; that 80 million, or about 85 per-cent, of these homes will be passed by a cable; and that 57 million, or about 60 percent, of all television households—or about 70 percent of the homes passed—will be cable television subscribers. After that point is reached at the turn of the decade, cable will grow as the population grows, but, since by 1995 most of the nation will have been wired that is ever going to be wired, any increase in penetration will be generated by market-ing rather than new construction efforts. In view of competition from other information and entertainment systems within the new electronic media, the most prudent estimate of cable sub-scriber penetration for the year 2000 is 85 percent. All of these consensus numbers, especially the numbers for the early 1990s, may be off target by two or three percentage points one way or the other, but the broad outline they present of cable television in American society is fact-based and reasonably accurate.

In digesting these numbers, it helps to recognize that most of New York, Los Angeles, Philadelphia, Boston, and Denver, and all of Baltimore, Cleveland, Chicago, Detroit, Milwaukee, Minneapolis–St. Paul, Sacramento, and St. Louis, are not wired and that on the whole they will not begin construction for another year or more. It will take five to seven years from the day construction begins, when the first worker digs the first hole or climbs the first utility pole, for these major cities to be fully wired. Cable penetration doubled in the last ten years, to a great extent because of new construction, and it will take another decade for it to double again.

At the same time, it is important to recognize that when we speak about cable, we are not speaking about a singular technical capacity or configuration. Today, 60 percent of the cable systems, serving 40 percent of the subscriber base, are 12-channel systems or less. Almost all the systems built between 1975 and the present are 36-channel systems. New systems being built today may have 50-plus channels, but in the early 1990s they will represent only about 15 percent of the homes passed by cable. The smaller, older-capacity systems are being rebuilt as their franchises come up for renewal, but after all is accounted for, the prototypical system in the early 1990s will be 36 channels, which is far cry from the 100- or 200-channel systems imagined by hyperactive journalists and science-fiction writers. The most important fact, however, is not how most people in America get television but what television they get and how they view it.

Pay television is often perceived as the dramatic opposite of broadcasting, since it is programmed with first-run movies without censorship or advertising, and supported directly by consumer subscription. Yet for consumers these distinctions have become irrelevant.

In the eyes of consumers, pay television is growing to become synonymous with cable television. Cable subscribers seem to view Home Box Office and the other pay channels as a fourth broadcast network. Today 60 percent of all cable subscribers are also pay television subscribers, and almost 30 percent of these subscribe to more than one pay service. In newly built and marketed systems almost all of the cable subscribers also subscribe to one of the pay services—leading one to the belief that by 1990 and beyond, 80 percent or more of cable households will also be pay subscribers, effectively erasing the distinction. This will become ever more true as channels such as the Playboy Channel and the Disney Channel offer a variety of programming that complements the foundation services in the middle, including HBO, Showtime, the Movie Channel, and Cinemax. Combining all of these, some pay-television experts are projecting a total of over 83 million pay subscriptions generated by the 57 million cable households expected in 1990. This suggests that the vast majority (80 percent) will subscribe to two separate pay services.

"I used to take my time washing the dishes and cleaning up after dinner, and then wander in and plunk down in front of the TV," a cable subscriber once told a researcher, "but now that I have cable, I look at the guide first to see if there's something I want to watch at eight o'clock." In many ways this woman is typical. Overall, people in pay-television households watch 17 percent more television than those without it, reflecting the greater promise of satisfaction their total television menu provides. But the promise of satisfaction comes from the total menu, not just the pay-television menu. In general, pay-television ratings over a year are almost identical to the average ratings for the third network in the broadcast sweepstakes. Most recently, that means that HBO's ratings and ABC's ratings are nearly identical. It is true that pay-television subscribers watch the broadcast networks only 85 percent as much as the nation as a whole, but when it comes to special events such as the Super Bowl, the Academy Awards, and beauty pageants, or to mini-series and made-for-TV movies, pay subscribers generate much higher ratings than other television households.

Cable subscribers no longer think of broadcasting and the networks. To them, HBO and Superstation WTBS, for example, are as much or more top-of-mind than ABC, CBS, and NBC. As cable subscribers scan their program guides or zap their remote tuners, the thirty-six channels are perceived as unequal only in the programming they offer. As Casey Stengel used to say: You can look it up. But when you do, you discover that viewers continue to harbor a fundamental satisfaction with the commercial broadcast networks. In the most competitive circumstances, the networks continue to capture over 53 percent of the viewing, and when they are at their best by everyone's standards, they do considerably better. It is only the observably weak programs that suffer in pay-cable households. This means that even if cable and pay television reached 100 percent of American homes, the three networks would still be viable. But currently and in the foreseeable future, with only 65 percent or ultimately 85 percent penetration for cable, the networks should be more than viable. They should be strong.

Broadcasters are discovering that they are a great deal better off than their own forecasts of doom (which they presented to the FCC and Congress) always implied. Advertisers have not abandoned them, nor have consumers. The key is programming, and broadcasters understand that they can adjust by tracking the differences in their programming schedules, when on one night they hold the attention of 93 percent of the total audience and on the following night they hold only 63 percent. Most important, they also realize that their advantage as broadcasters is that they can reach all of the people, and although they may not do so *all* of the time, no other medium in the foreseeable future will reach all of the people at *any* time.

The Rise of Ted Turner

The current wisdom is that Ted Turner was born right on time, but for many years Turner believed that he was born too late. When he entered broadcasting with a UHF station license in Atlanta, Bill Paley, in Ted's personalized way of thinking, already owned the ball, the arena, the players, the referees, and the Astroturf. It is hard to start a railroad—to use another analogy—when the previous barons own all the rights-of-way. In the combination of the satellite and cable television Turner saw the opportunity to make room for himself to become the "fourth network" mogul. Turner spoke of creating the fourth network at the same time that he called the networks dinosaurs. Neither statement was quite true. Superstation WTBS is only the "fourth network" if qualified by the words "broadcasting" or "commercial." In terms of historical precedence or viewership, Home Box Office is the fourth network. And the networks are not dinosaurs facing extinction, particularly since cable television distributes the programming of the three commercial networks and their affiliates.

Nevertheless, the combination of Ted Turner and the launch of his Atlanta station on the satellite in 1976 symbolized that the rules had changed, but not entirely. WTBS is after all a broadcast station, wholly supported by advertising dollars, and providing the same programming as any major independent station in a major market like New York, Chicago, or Los Angeles—a combination of sports, movies, and reruns of old network program series. Cable operators pay Southern Satellite Systems, the common carrier that distributes WTBS, a fee of ten cents a subscriber each month, but Turner sees none of this money. His income is from advertising. He is simply an exporter of what cable systems had always called imported stations. Turner's first major act of adroit salesmanship was to persuade cable operators to make WTBS their primary imported station.

Independent television station audiences were growing and WTBS served these audiences in cable households. Viewers who grew up with television were inclined to watch independent television stations, sometimes as much as ten or more hours a week, or about one-third of the time spent watching network programming. The ratings reflected this one-third cut enjoyed by strong independents. If WTBS became America's independent in cable households, it should perform accordingly; and it did. The evidence for the future is compellingly clear. Americans have consistently devoted a measurable amount of their viewing loyalty to the kind of programming found on independent broadcasting stations, first in those markets where major independent stations existed and then in cable households where the importation of WTBS and other broadcast stations made Americans equal. As cable grows, this will continue to be true for the foreseeable future. These basic facts create the first foundation for understanding consumer viewing behavior in the decade ahead.

If Ted Turner had only imagined and promoted WTBS as an exported Superstation, he would only be remembered as a footnote in the history of cable television and as one of the more colorful personalities of our times. Ted Turner is one of the people whose personal history coincides with the history of Cable Television, and he looms as an important factor in the future of communications because he created the Cable News Network. Although WTBS went on the satellite in December 1976, it belonged to the previous history of cable's importation of independent stations. The Cable News Network, however, was not only part of cable's third and current postsatellite historical age, but, along with Home Box Office, it defined the age with its birth, on June 1, 1980. It also changed our understanding of news.

By this time, the Madison Square Garden Network had changed its name to the USA Network because it had filled out its prime-time professional sports programming with contracts with the baseball leagues, the National Hockey League, and the National Basketball Association. Before the games, it programmed "Calliope: Entertainment for Growing Minds," a series of distinguished children's films, and in September 1979 it began to devote the remainder of its time to live coverage of the House of Representatives. This programming is now on C-Span, and the USA Network expanded its schedule with women's service programs and movies. But the network has never acquired a distinctive image.

ESPN, the twenty-four-hour sports channel started by the Rasmussen brothers in September 1979, and then sold to Getty Oil, had difficulty obtaining programming (one viewer described it as "watching frogs jump in Indiana") and untangling its business. Warner Communications had begun Nickelodeon that same year under the guidance of Dr. Vivian Horner, but it was not perceived as big-league both because it did not accept advertising and because it offered educational children's programming. There were also some less inspired networks on the satellite little known outside the cable industry. Thus when CNN came along, no satellite cable network had carved its niche in the public imagination nor in the communication industry's conceptualization of the medium.

The idea of a twenty-four-hour news channel was in the air. The way had been paved by the Associated Press, Reuters, and the United Press International, all of which had provided automated news ticker wires to the cable industry for over a decade. UPI had added slow-scan pictures in July 1978, with an audio track from its radio service. Cable operators knew that they got more subscriber complaint calls when these services had technical difficulties than they did for any other service or channel, but alphanumerics and slow-scan pictures were, after all, not the "real thing." Time Inc., *The New York Times,* and the Post-Newsweek Stations, all with extraordinary news credentials, were openly exploring the concept, but backed off because they thought the plan uneconomical and impractical in logistics. They also predicted Turner would fail for lack of funds and journalistic credentials, in which case the field would be left open for their later entry.

They had been warned by industry consultants that Turner would not fail. As Daniel Schorr had pointed out, "Bill Paley, when he founded CBS News, had made a great deal of money in entertainment, and his path just happened to intersect with that of a man named Edward R. Murrow." In Turner's case, his intentions intersected with the available talent of Reese Schonfeld. The titans of existing journalism had also been warned that the economics would work in Turner's favor and that if they didn't enter, the field would be closed for eight years or more; there would simply not be a sufficient subscriber base for a second network until then. In any case, the field was anxious and alert and open. Today not only is CNN a success, but the failure of the competitively launched Satellite News Channel, backed by mammoth Group W in 1982, proved the critical mass theory that there wouldn't be room for a competitor yet and there probably wouldn't be until the turn of the decade.

Part of the revolution in television news was based on the concept of video publishing. In trying to distinguish himself from advertising-supported news programs carried on broadcast networks, Turner once exploded, "They don't tell you Chrysler's going under, because they don't want to lose the car ads. They don't investigate DC-10 engines falling off because they don't want to lose the airline ads. But we're not dependent on advertising, because half our income is the fifteen cents a viewer pays each month to watch us. Don't any of you news guys understand the way Cable News Network works?" Although Turner's examples were exaggerated, his analysis of the differences was right on target. Subscriber-supported cable programming can have an editorial voice. It can also air programs twenty-four hours a day even though doing so vitiates the ratings and the size of the advertising revenue.

Furthermore, twenty-four hours of continuous live video coverage of the world is a breakthrough concept in its own right. It can show you the demonstrators going home after the other broadcast television cameras are shut off. It can follow a slow-developing story hour by hour and day by day to its denouement. It is participatory and available when you are, and the never-ending dynamic of the channel enables it to play antic havoc with tradition-bound news coverage. The simple availability of the channel changes the rhythm of political action and changes our personal horizons. Like no other enterprise before or since, the Cable News Network intrinsically articulated what cable television is and can be.

Allocating Cable Channels

There are thirty cable satellite networks at present count, but they are by no means equal, nor do they provide equal glimpses into the future. These networks can be organized into three groups. The first group reaches 50 percent to 90 percent of current cable subscribers; the second reaches 20 percent to 50 percent; and the third reaches 15 percent or less. In the first group are ESPN, WTBS, CNN, USA, CBN, MTV, Lifetime, Nickelodeon, TNN, and C-Span. As a group they represent sports, news, movies, contemporary music, children's, and general independent-station–like programming. The second group includes the Financial News Network, the Weather Channel, Arts & Entertainment, Black Entertainment Television, and other more specific-interest networks. The third group contains religious, ethnic, and educational networks. The criteria for these groupings are cable system channel capacities and the marketability of the programming content. Together they represent a reasonable picture of the future.

If we accept our earlier reasoning that the prototypical cable system in 1990 will have thirty-six channels, it would be realistic to examine how a cable operator might allocate this capacity according to his own marketing needs. Using a current major market system as an example, we discover that thirteen channels must be dedicated to the broadcast stations viewed in the area. Six channels are dedicated to local origination, public access, and government use as required by the franchise agreement. Three channels are automated, providing racing and stockmarket results, a program guide, weather, and a community bulletin board. One channel is dedicated to C-Span, which covers congressional hearings. Three channels are dedicated to pay television. This leaves only ten channels available for the other satellite networks. The spread for more openings than that occurs only if there are fewer broadcast channels in the market or if the system has more than thirty-six channels.

The cable operator's allocation of channel capacity is analogous to the allocation of retail floor space in a department store or shelf space in a supermarket. Each channel must make a direct contribution to the overall service and a contribution to profits. Programming is a marketing instrument and thus cable programmers will choose channels based on the demographics and psychographics of their market, for example, choosing the Disney Channel over the Playboy Channel. Operators will also favor pay channels that generate direct income over satellite channels that they must pay for. But this cannot be done at the expense of broad market penetration for the service as a whole, and thus channels such as C-Span and Nickelodeon will be included because of their obvious value for citizenship and children. In choosing a basic mix of channels, therefore, the cable operator will favor consumer pay broad-scale channels for reason of economic support and programs dealing with children, news, weather, health, religious, and ethnic programming for social and political reasons.

When all of this classical, business, and street thinking are put together, it becomes clear that the ten largest cable satellite networks today represent the basic substance of what we can expect in the future. Of course, there will be other channels, some reaching 40 percent of the cable universe, and there will be directly competitive channels, such as a second music channel or news channel, and there will be improvements in the programming of all the channels. But in general, it is not unfair to say that what you see is what you'll get, at least in 1990.

If these speculations about the future of cable programming are reasonably accurate, it is also possible to make a reasonable judgment of consumer viewing patterns in the future based on the viewing patterns we can now observe in pay-television households on thirty-six-channel systems. Channel capacity is critical, for as it grows from four to thirty-six channels, network broadcasting's share of viewing drops from 90 percent to 56 percent. In pay-television households the network share drops to 50 percent. But both channel capacity and pay television increase television *usage* with the result that though the broadcast networks may lose shares of the audience viewing, this may not necessarily be reflected in the ratings or in the actual number of viewers.

Based on all the available research, then, it is reasonable to suggest that the typical cable household in 1990 will be watching sixty hours each week; that 50 percent of this viewing will be to the broadcast networks; 15 percent will be to independent local stations and to public television; 15 percent will be to the various cable networks; and 20 percent will be to the pay-television networks. This also means that although some cable networks will outperform others, the average rating for a cable network program will be 1.5 percent, or about 850,000 viewers in 1990. It means that in cable households, pay-television programming will be outperforming one of the three broadcast networks.

In this general survey of cable television's past, present, and foreseeable future, we have tried to come to an intrinsic understanding of what can be fairly called "cable television." Three subjects remain, however, that have clouded current understanding: other satellite applications, broadband communications, and local origination.

The three satellite applications are Direct Broadcast Satellites (DBS), Satellite Master Antenna Television systems (SMATV), and pay-per-view events. DBS and SMATV stimulate dreams of our own personal earth stations roaming the heavens like vacuum cleaners sucking in all the programming on all the satellites in geosynchronous orbit. The fact of the matter is that DBS, which does envision an earth station in your backyard, and SMATV, which envisions an earth station on the top of high-rise apartment buildings, can neither technically nor economically provide more channels than cable television. The most powerful DBS station proposes only six additional channels over broadcasting, and SMATV is nothing but a mini–cable system serving one or several buildings, rather than a town, village, or city. Thus DBS is really cable television for rural areas where stringing cable would be uneconomical, and SMATV is only a way for specialized groups, such as upwardly mobile singles or senior citizens living in particular apartment clusters, to have a cable system more attuned to their preferred needs than they might expect from a system serving the community at large. For our purposes of envisioning the communications map of the future, they simply fill in pockets of the topography with different distribution systems for the same product as cable television.

Our dreams of earth stations in the backyards, however, do not end with such clearly defined businesses as DBS and SMATV. Today there are over one million satellite receiving dishes in American yards, and their numbers are growing at a rate of 50,000 a month. Essentially, these people are thieves and peeping Toms who access not only the cable and pay networks without paying for them as their neighbors do, but also foreign programming, broadcast network feeds, and space shuttle conversations. On the grounds of privacy and just payment, these communications will eventually be encoded or scrambled. It is a technical and regulatory nightmare, but at the rate of growth of backyard dishes, encoding is inevitable. The result will be that you will only be able to get what you pay for, which should not alter our general view of the communications map of the future.

Videotex and other broadband services will be discussed in a later chapter. What is important for us at this point is to recognize that broadband services need not be delivered by the wires of the cable system, and if they are provided by the cable system, we should probably change our vocabulary to keep our conception of the services distinct. *Cable television* is a useful term for the broad-scale distribution of broad-scale programming, which is a different concept, product, and service from the interconnection of computer terminals, even if provided by the same company over the same wires.

Local origination programming, on the other hand, is a concept intrinsic to cable television and will undoubtedly undergo considerable development in the coming decade. Amid all the facts and fantasies about cable television, we have overlooked its potential relationship to newspapers and the kinds of communications provided by newspapers.

To date, newspapers have been the closest medium to shaking hands with your neighbors and being invited in for tea.

They are Main Street, as opposed to the Interstate. They can be delivered to your door and enjoyed with coffee. They are read on the train to work and in the barber shop. Newspapers cause conversation—at the post office, the bank, the beauty parlor, the supermarket, and over the phone. They tell you the price of pork chops and automobiles; who died and who got married; what jobs are available and what houses are for sale; what's on television and at the movies; your future under the stars and the score of last night's baseball game under the lights. A newspaper is the mass medium of its community, the grounding of commerce, opinion, and decision. That is why 75 percent of all adults collectively spend over $10 million each day to read a newspaper, indicating that they consider a newspaper critical for their involvement in the life of their community. That is also why newspapers are the largest of all advertising media, with revenues of almost $15 billion, or 29 percent of all advertising media investments distributed between 2,400 local daily and Sunday newspapers.

After an analysis of the most recent research data, the Newspaper Advertising Bureau reported that among individual newspaper sections, general news drew the highest readership (94 percent) and four sections—editorial, entertainment, sports, and TV-radio listings—shared second place with 81 percent of the total weekday audience. Readership of other sections was reported as follows: comics, 79 percent; classified, 78 percent; food and cooking, 77 percent; home furnishings and improvement, 76 percent; and business and finance, 55 percent. These statistics tell us what people actually do with their newspapers and what they expect to find in them.

Cable television, however, is also a community-based medium more than capable of reporting the shenanigans of the city council and the wedding of the widow and lifelong bachelor in the retirement home. Certainly cable through local origination channels can advertise the price of pork chops and the values to be found in the community boutique. Absorbing newspaper's advantage, cable is also able to inform you about yourself or the ball game you were at the night before, or to repeat the high school football game or local Veterans Day parade while making editorial comment on the proceedings. Community bulletin boards and television program listings are becoming a staple of multichannel systems. It's possible the cable operator or the retail advertiser may find an enlightened way to put display advertising on cable channels that may affect the department store and supermarket advertising currently found in newspapers.

At present, cable operators are required by local franchise agreements to include local origination channels, but these are not yet functioning at their peak. If their potential were in the hands of a community-minded entrepreneur, cable could have the same impact on local newspapers that broadcast television had on *Life, Look,* and *The Saturday Evening Post.* Consider the effect if the satellite-delivered channels included a greater proportion of service editorial for homemakers, stock brokers, sports enthusiasts, and all concerned about health, wealth, happiness, and relationships. While a community must exist before there can be community television, it is equally true that cable television has the capacity to create community. Probably the next significant wave of cable development will be to discover these roots and complete the definition of what cable is and can be.

The Way of Technology

History is filled with stories of people who have tried to force technology to serve their ends and of others who have been afraid to use technology. These stories have repeated themselves in the history of cable television. Once viewed as a medium favoring the receiver, cable television is now seen as a sender's medium. Once seen as a hardware business, cable is now seen as a software business. Once thought of as a public utility, cable is now considered a vehicle of video publishing.

And yet private dreams of private television continue to run rampant in the press.

These dreams see cable transforming television into a service like the telephone, providing the consumer with nearly random access to programming so direct, so specific to personal tastes that its attractiveness to others is nearly irrelevant, almost as irrelevant as what the rest of the world might think of what a friend or relative says to you at the end of a telephone wire. But until the world is wired with fiber optics, and cable and telephone companies merge to provide a radically new kind of service with a different name, cable television will remain a medium of distribution for relatively broad-scale television programming. Today you can read twenty-two different magazines on fishing and thirteen on motorcycling. Several articles in every issue of each of these magazines are the equivalent of a television program. At no time will cable television provide this kind of specialization and nearly random access.

Technology ultimately has its own way; for, like any tool, art, or craft, it refuses to be used against its nature, and over time people will use it correctly. What we call *cable television* originally called itself *community antenna television*. It changed its name once, and it is quite possible that it will again. At the moment, however, the current technology offers us a community-based medium for the rather broad-scale distribution of television programming based on news, entertainment, and service. It is a medium for everybody in the community. That concept, that product, will endure whatever name it uses, and we should plug part of our vision of the future into its athletic imagination.

5

Personal-Use Media

On February 27, 1966, *New York Times* columnist Jack Gould reported that Dr. Peter Goldmark of CBS Laboratories had developed "a metal disc that reproduced motion pictures through a television set in much the same manner as a long-playing record reproduces music through a high-fidelity phonograph." CBS officials vigorously denied that the story had any foundation in fact, but Dr. Goldmark's reputation for electronic wizardry, coupled with Jack Gould's own journalistic prominence, convinced many people that not only a new invention, but possibly a new age of communications was about to be born: the age of personal choice television.

Consumers today purchase millions of videocassettes, videodiscs, video games, and home computers—just as they purchase millions of television sets. Consumers *use* these devices in very different ways, based on individual, personal choice, as with records and books. The information and entertainments these devices offer—be they children's games or educational programs, home videotapes of a child's graduation, first-run movies, or spreadsheets for home banking—represent competition for the broad-based media. Personal-use electronic programs of every kind assume a community of people interested in them, though these communities are counted in thousands or hundreds of thousands rather than millions or tens of millions. Potentially they segment electronic communications into ever smaller and more disparate groups. More important, however, even if everyone purchased and ran the same program, no two people would ever be doing it at the same time. They would be doing it on different days, at different hours, under different circumstances, with isolated inputs and reactions. This individualization of time and experience is at the root of what the consumer dreams about and does with these new media.

The personal-use media seem to have been born in newsprint and nurtured by speculation. An imaginative public has consistently accepted "future shock" stories about these new media even before laboratory prototypes, marketable machinery, and programming were available. This turmoil arises in part from electronic manufacturers' battling for position in the public mind, marketing by press release to preempt competition and establish their particular technology as the industry standard. It also arises out of the public's delight in the glamour of gadgetry—the smoked glass, chrome, and fins of technology. And then, too, the exciting new media all seemed so possible. American and Japanese engineers could make anything they imagined, and both video recording and computers seemed ready for consumer exploitation. But beneath it all was a simple truth: personal-use electronic media were an idea whose time had come.

What Had Goldmark Invented?

It was easy to believe that Dr. Goldmark had developed a videodisc, because as early as 1927 British TV pioneer John Logie Baird had demonstrated a crude video picture from a form of videodisc based on the Edison gramophone concept, and as recently as the previous July 8, 1965, CBS had used a magnetic disc for the first instant replay during a football game. Certainly Goldmark, who had invented the long-playing record and the color television system used on our moon voyages, could produce a workable consumer videodisc. In late summer of 1967, CBS finally officially admitted that Dr. Goldmark had in fact developed an Electronic Video Recording (EVR) system, and a full year later, on October 10, 1968, CBS demonstrated a prototype monochrome player in New York City.

What Goldmark had invented was a high-density information storage retrieval machine, a more compact, higher-resolution, animated microfiche. Through a misunderstanding possibly based on an inventor's hubris and an entertainment conglomerate's corporate politics, EVR was presented as a computer product for the playback of movies in the home, with the added advantage of instructional programming. But movies—the visual equivalent of novels—are relatively long, narrative, and discursive in style; their intrinsic art is the direct opposite of data retrieval or high-density programming. In effect, EVR was far too expensive to use for viewing movies. The key thing that went wrong with EVR from the beginning was lack of understanding of what had been invented.

In Kantian terms, there was a lack of understanding of the *Ding an sich,* the thing in itself. In many ways the failure of EVR was actually a triumph of technology over promotion, for the technology refused to be twisted into a consumer home-movie machine. The failure to understand and accept the *Ding an sich* of technology has plagued the consumer electronics industry since the announcement of EVR, inhibiting both the appreciation and marketing of videocassettes and videodiscs and repeatedly confusing our interpretation of video games and home computers. We will look more closely at this confusion and unravel its twisted threads more carefully later on.

EVR turned out to be something quite different from a videodisc. The technology used a narrow strip of nonsprocketed film read by a flying spot scanner. The film was encased in a self-contained seven-inch cartridge and provided fifty minutes of monochrome programming. The player unit could be attached to the external antenna terminals on the rear of any television set and provide home viewers with their choice of television programming at their convenience. Dr. Goldmark had not precisely produced the system everyone had had in mind, but he certainly had given life to a concept that fired imaginations.

Then, just before Christmas on December 22, 1971, CBS announced its withdrawal from EVR involvement. In any business mistakes will be made, but the basic problem in this case was that too many of the participants had tried to imagine into reality a product that did not exist in the current technology. People wanted something to exist that simply did not exist.

On January 24, 1972, the day Dr. Goldmark's resignation from CBS was announced, Sears, Roebuck ran advertisements in Chicago newspapers announcing that the first videocassette consoles would be on sale the following June in eighteen Chicago area stores. Nearly four years earlier, brothers Arthur and Frank Stanton, enormously successful Volkswagen dealers, had visited the booth of Playtape, Inc., at the 1968 Consumer Electronics Show in Chicago. Playtape had developed a system for cramming a lot more programming onto a significantly smaller amount of tape than had been possible with previous videotape technologies; the Stantons saw in this new technology the opportunity of bringing videotape within the range of the consumer's pocketbook. They purchased the company and set about seeking an arrangement with a major electronics manufacturer who could provide electronic expertise, financial backing, and manufacturing facilities to develop their product. By mid-1969 they had sold a majority interest to Avco Manufacturing, and had changed their company's name to Cartridge Television, Inc., and their brand name to Cartrivision. Cartrivision quickly became the most vigorous company in the video player sweepstakes. But, more important, Cartrivision defined the product for the industry—and the definition was greatly misleading.

The problem with VCRs as a consumer product was their need for large quantities of videotape, which made the cost prohibitive for most consumers. This was in part a technical problem, and Cartrivision tried to solve it—but at the cost of quality, and with the need for an integrated and expensive console. On another level, the problem was the way the product was conceived: It was considered "software led," which meant that it would be used for prepackaged programming that consumers hopefully would flock to as an alternative to the homogeneous programming of American broadcasting. In simplest terms, videocassettes were thought to be analogous to audio records and tapes, which were independent of radio. Few believed that consumers would use their VCRs for recording regular programs off the air.

When this belief in the product's dependence on prepackaged software encountered the expensive realities of videotape economics, Cartridge Television opted for a rental strategy to make VCRs more acceptable to consumers. It set about obtaining Hollywood films by assuring producers that the cartridges would be "gimmicked" so they could be played only once. At the same time, it designed a franchise distribution system based on movie theaters. RCA tried to conceptualize a similar relationship with Fotomat with the same goal in mind, but the very idea frightened most hardware manufacturers who did not have interests in Hollywood. In 1973, after delivering only six thousand machines, Cartridge Television threw in the towel and declared bankruptcy. By this time, however, most of the consumer electronics industry had concluded that prepackaged programming was the key to success and a much less expensive technology (to say nothing about one that worked) had to be found. RCA, in particular, concluded that that meant a video disc.

From the time of EVR's announcement to the bankruptcy of Cartrivision, there was as much or more promotion and excitement generated by videocassettes and the "new video" as there is today over home computers. Perhaps this is a cautionary tale in itself, but to understand the times it is important to recognize that there had been a decade of failure by major companies who had promised new video for the home of the future. To understand the past and to project the future, however, it is useful to focus on people rather than technology. The two key people, representing a clash of East-West cultures and offering opposite insights into technology, were the ghost of David Sarnoff, who haunted not only RCA but the American consumer electronics industry, and the living genius of Akio Morita, the founder of Sony.

In many ways, RCA and David Sarnoff were the consumer electronics business in America. RCA had struggled with consumer videotape technology, and after years of development admitted failure by pulling back from its highly publicized MagTape system in 1974. RCA being RCA supported the American assumption that it couldn't be done. But RCA never believed in videotape, for General Sarnoff had imagined discs, not tape, and although the company kept up its competitive front with MagTape, it was convinced that videotape was a product for the future, to be used in conjunction with new breakthroughs in television sets and distribution technologies.

With an imaginative and technical aggressiveness that a young David Sarnoff would have understood, Morita insisted on tape, supported both by his own engineering talent and by the Japanese passion for recording. Next to RCA, Sony was a small company, and in the United States, for all its reputation, had less than a 7 percent share of the market. But Morita had seen his half-inch reel-to-reel videotape recorders grow to dominate the school and corporate markets after failing as a consumer product in 1965. In November 1969, Sony demonstrated a "color video phonograph," a rather clumsy expression, but that seemed to be what you had to say in those days. It also coined the term *videocassette*. These machines went on sale at Neiman-Marcus for $1,300 in February 1972, using three-quarter-inch tape and called U-matics, but they never became a consumer product. Instead they became the lucrative standard for almost all other applications, from portable news cameras to corporate training programs.

In the spring of 1974, in what most considered a publicity stunt by Sony to hold back the growing support for RCA videodiscs, Sony gave birth to MAVICA, and the VCR technology took a giant step closer to becoming a popular consumer product. As initially demonstrated, MAVICA (for Magnetic Video Card) was a flat seven-by-nine-inch card of brown magnetic material inserted into playback equipment, almost exactly as the RCA videodisc was. It could hold only ten minutes of program material, but when the technology was applied to half-inch videotape, it reduced the raw cost of the tape and cassette required for a one-hour program to less than $15, and soon to half that. Given the brand name Betamax in its cassette form, the product was introduced in a console model, with a nineteen-inch Trinitron color television set, in 1975 for $2,295. The following year, a tape deck alone was available for $1,300. The consumer VCR market had been born.

The VCR Revolution

Engineering breakthroughs and price points were certainly important, but the real revolution was that Morita successfully introduced the hardware without software, with a single-minded belief in the technology itself. As a thing in itself, the videocassette recorder is simply a device for recording and playing back television. With VCRs you can record a program, view it, review it again as many times as you wish, erase it, and record something else. Morita's VCR was never the least bit dependent on prepackaged programming for sale or rent. In fact, the technology resists this use. Morita's mistake was his insistence that one hour of recording ability was all that was required to satisfy consumers. For this reason alone, Sony rapidly lost its dominance in an industry it had created.

If videocassettes represent a cultural revolution for the consumer who can now record and play back television programs, Betamax caused a second revolution in the consumer electronics industry. Having admitted failure in tape development, the macho American industry surprisingly and quickly aligned itself with the Japanese, to the extent that all VCRs now sold in the United States are of Japanese manufacture. RCA aligned itself with Matsushita, the second-largest television manufacturer in the world and the parent company of JVC, Panasonic, and Qasar. JVC had developed VHS (Video Home System) with a longer recording time than Betamax. RCA marshaled its marketing clout, while Sony failed to increase recording time. The result was that the VHS format almost immediately captured 60 percent of the market. The VHS format now has 75 percent of the total market, which has boomed beyond most expectations.

In 1983 there were 4.1 million VCR units sold in the United States, more than doubling those sold the previous year. In 1984 sales nearly doubled again, to 7.6 million. In 1985 almost 12 million VCRs were sold; VCRs can now be found in 31 percent of American homes. These sales figures almost exactly replicate the first boom years of color-television-set sales. They establish not only a clear growth curve, but high expectations. Based on these kinds of sales figures, it is easy to imagine a VCR in 45 percent to 50 percent of U.S. households in 1990, and 75 percent to 80 percent in the year 2000—keeping in mind that color television sets are in less than 90 percent of our homes today.

These current projections assume a reasonably healthy economy, moderately falling prices, and the general absence of competition, which could include videodiscs and home computers as well as dramatic breakthroughs in television sets themselves, such as changing the size and resolution of the picture and adding some built-in memory device. The impact of such competition on the VCR growth curve, however, will not emerge for several years. Thus even the most conservative projections hover around 40 percent penetration at the turn of the decade. In cultural terms, the most conservative projections imply that those in affluent, educated, style-conscious, trend-setting sectors of the population will all own VCRs and that their children will grow up with VCRs as much a part of the household as television itself. Understanding what people do with their VCRs is thus to understand an important part of our future values and behavior regarding communications.

All the studies and surveys of VCR usage support our perception of the technology's *Ding an sich*. The typical VCR is used around thirty times each month, 40 percent of the time for recording, 60 percent for playback, with people watching what they have recorded more than once. Only 30 percent of the playback time is used for rented or purchased prerecorded material, and a significant percentage of this material would be described as X-rated. Only 10 percent of VCR owners have a home television camera, and less than 5 percent of their viewing is to what is generally regarded as home movies. There is no question today that the primary use of the equipment is for recording televised programs such as network and pay-television movies, soap operas, sports, and popular series such as "Dallas" and "Dynasty."

The videocassette industry has come to call this phenomenon "time-shift," and the vast majority of people do not store the programs for very long. Interestingly, VCR owners do not report watching less broadcast programming. Rather they increase their total time with regular television, using their VCR as a tool to increase their options and create their own time-frame. This dynamic, rooted in the intrinsic characteristics of the technology, will survive all competition and contribute to an increase in communications consumption and to dramatic changes in the way we schedule that consumption.

Renting was never an American consumer habit; but, to the surprise of many, the overnight rental of videocassettes has become a significant business. In 1985 revenues from the sale and rental of videocassettes went over $3 billion, rapidly closing on the revenues generated by movie theaters. About seven out of ten VCR owners rent cassettes, averaging thirty-five rentals a year. Over 70 percent of all movies produced in the last fifty years are now available for home rentals. Yet in spite of the rentals, the primary use of VCR is for recording and viewing programs that could otherwise be seen on broadcast or cable television. TV consumption has increased, but the dramatic difference is the privatization of time.

Videodiscs: The Elusive Dream

At a seminar of business executives in 1980, Dave Lachenbruch, the editorial director of *Television Digest* and *Consumer Electronics,* and without doubt the most knowledgeable observer of the industry for thirty-five years, answered a question that had been repeated thousands of times since the first discussions of personal-use home video almost fifteen years earlier. The question was: "When will I be able to record on a videodisc?" With the look of a war-weary veteran, Lachenbruch replied: "When you can record on a videodisc, it will be a videotape." He knew that there had been experimental models of recordable videodiscs using optical and thermoplastic techniques, but he went to the heart of what we have been calling the *Ding an sich*. There is a fundamental difference between a playback-only device and a record-playback-rerecord device.

On June 24, 1970, in Berlin, four days before the first Cartridge Television demonstration in New York, and five months before the first delivery of an EVR machine, the Teldec videodisc was demonstrated. A joint product of German and British scientists, the flat flexible disc carried five minutes of monochrome television images viewable on a standard television set. There were obvious problems in the system's limited program capacity and lack of color, but the product unveiled at that moment articulated and became the imaginative standard for everyone who contemplated a distinctive programming-based consumer market. The idea was possible.

Teldec equipment eventually reached the West German market in March 1975, but the pent-up consumer demand to be liberated from government-controlled broadcasting was satisfied by the simultaneous availability of VCRs. In fact, Europeans currently are buying VCRs at a much more rapid pace than Americans are. When the General Corporation of Japan introduced a jukebox enclosing fifty Teldec discs in 1980, it seemed to announce that both the market and the technology were obsolete.

Unquestionably the most dramatic demonstration of electronic hardware I have seen in the last twenty years was N. V. Philips's preemptive strike against the Teldec videodisc at an international video conference in Cannes, France, in 1972. The demonstration took place in the theater of the Palais de Festival, which was equipped with a large projection screen framed by proportionately large digital readout displays, much like the recent electronic scoreboards. Television monitors were placed at strategic locations to show the picture quality always lost in projection. The equipment being demonstrated was an optical videodisc, with thousands of individual frames read by a laser beam that defied wear-out and had pinpoint accuracy. The quality was staggering, but the real impression was made by the equipment's versatility, its ability to pick any of 50,000 frames at random or in instantaneous complicated sequences. The still frames could be displayed indefinitely or played at any speed, forward or backward.

It was a magic show, the first new video invention that made you want to create something—something interactive, something crazy, maybe a new kind of whodunit movie. The imagination soared and dreamed, and then came to rest. The system was expensive and nobody knew what to make of it. The machine would probably be more expensive than VCRs for a long time. If it was simply a device for viewing movies, consumers would probably not buy it. At comparable prices they would select a product that could record and reuse other programming rather than this one whose main feature was search-and-discover within a program. No one really knew how search-and-discover possibilities would be used by consumers in a home situation.

The first videodisc equipment went on sale for test-marketing purposes at Rich's Department Store in Atlanta in December 1978, and people lined up and waited overnight to buy the first thirty players at $695 each. But since then there has not been a great deal of good news for the product. The prices have increased over the $700 barrier, while VCR prices have decreased. There is nothing terribly special about the programming or its pricing, and there has been no unified or salient marketing effort. Less than 500,000 optical videodisc players have been sold by Magnavox and Pioneer since their introduction. Yet the dream lives on, especially when thought of in conjunction with a home computer.

From the beginning, RCA had its heart and mind set on videodiscs, and throughout all the public turmoil of various systems, throughout multiple management changes, and even during the early success it enjoyed with VHS recorders from Japan, its focus remained remarkably consistent. RCA always thought that a low-priced videodisc system—the magic number has always been half the cost of a VCR—would be the next color television for the consumer electronics industry. Color television was RCA's template, a hugely successful product inspired by the vision of the singular David Sarnoff. NBC financed color programming beginning with "Walt Disney's Wonderful World of Color" and then "Bonanza," as well as parades and sporting events. RCA built and installed broadcast transmitters for NBC stations and then for others. And RCA, Indianapolis, was for a long time the only manufacturer of color television sets.

That time, in fact, was longer than most people remember. The consumer electronics industry was not anxious for the intrusion of color TV, seeing no reason to threaten or confuse the market with a new product while business was doing so well with black-and-white sets. At the same time broadcasters in general had little motivation to invest in more-expensive color programs and transmission facilities while there were so few homes with color sets in their audience. Until color television programs generated larger audiences, color programming was simply an unnecessary expense. But General Sarnoff kept manufacturing and programming for color for nearly ten years. Suddenly, in 1965, everyone came aboard, and color set sales jumped from hundreds of thousands to multiple millions per year.

RCA imagined the same success for videodiscs with RCA Consumer Electronics manufacturing and retailing the players and with NBC and RCA Records developing and distributing the discs. What they failed to account for in their thinking was that every color program the NBC Network had invested in could also be seen in black and white on regular sets, thus returning the programming investment, and that every color set RCA sold at a premium price could also receive the full menu of black-and-white programming. Color television sets did not require the industry unity a videodisc might need to sustain a long gestation period.

RCA announced its videodisc system in June 1972, and demonstrated it widely in the early 1970s in an effort to persuade all the relevant industries to join in a united effort. The RCA technology was built around a capacitor, the same technology most radios use for station tuning, with a stylus riding in guidance grooves over the disc's pitted surface, the depth of each pit determining its signal information. Since the pitted disc is subject to dust and fingerprints that could distort the picture, it is encapsulated in a plastic sleeve. The sleeve, like a record-album-size credit card, is inserted in the front of the player, where the record is released, played, and returned to its sleeve.

Next to an optical videodisc, this capacitance system (CED) was simple and pedestrian, and many in the industry did not feel that the mass market needed another signal source for movies. They felt that broadcasting cable, pay TV, and VCRs were sufficient. RCA signed production and distribution agreements with other manufacturers and retailers, but essentially the company went it alone when it finally put the players and records on sale on March 22, 1981. In ten weeks, dealers had received over 60,000 players, and by Christmas consumers had purchased 100,000 units. It was the most successful consumer-electronic launch in history. It had taken color five years and VCRs two and a half years to accumulate the sales that the videodisc achieved in little more than six months. Through some perversity, energized by RCA's passionate promotion of higher expectations, the press called this success a failure. In 1982, 223,000 players were sold and in 1983 350,000—a total that VCRs took four years to match. Meanwhile, the discs themselves offered the same programming for sale that VCR owners were renting.

There can be little secret about what people do with videodisc players in their homes because the equipment does only one thing, but the quantity of videodisc activity could be perceived as astonishing. On the average, including both new and veteran owners, people purchase thirty discs per year, and they use their players fifteen or more times each month. This experience is being increasingly reflected in videocassette sales. Almost 70 percent of VCR owners now purchase at least one full-length movie a year at prices as high as $79.95. Most of the growth, however, is coming from the purchase of unique, repeatable programming, covering the spectrum from Jane Fonda's *Workout,* which has sold a half-million copies, to children's programs—such as Richard Scarry's *Nursery Tales*—which now represent over 15 percent of all cassette sales.

During the first week of April 1984, RCA's board of directors voted to halt manufacture of videodisc players. This brought to an abrupt end a fifteen-year and almost $800 million effort to establish the technology. RCA had tried to go it alone as they had with color television, but they found themselves all alone in the industry longer than they could afford. While other manufacturers, record companies, and Hollywood film distributors waited for the player population to grow, prices dropped to less than $200. At the same time, the RCA board was looking at revised estimates for VCR sales, which suggested the movement of 6.7 million units in 1984 and consumer expenditures reaching $4.5 billion. Whether Japanese manufacturers continue to sell CED players is unknown at this point, and few expect sales of optical systems to take up the slack and become a mass-market product in the near future. Even if we concluded that the consumer had yet to cast a definitive vote, certainly the consumer electronics industry has voted that VCRs are the product consumers will have.

Cassette rentals currently run between one and four dollars per day, and the number of video stores offering convenient service is exploding. The cost to a film company to reproduce, package, wrap, and ship a videocassette is about $12, and declining. Generally, this provides a margin of $13 in the wholesale price for the film company's talent expenses and profit. In this context, it is becoming increasingly easy for film companies to earn $5 million to $6 million in profits from the cassette distribution of a film. This means that films will continue to be distributed on cassettes. It also means that in the future VCR owners will probably behave like disc owners, renting and buying up to forty cassettes a year without diminishing their record-playback behavior.

The Lessons of Video Games

While the titans of the consumer electronics industry were battling over videocassette and videodisc markets and technologies, three young inventor-entrepreneurs in California's Silicon Valley developed products that challenged and transformed the home entertainment industry. Nolan Bushnell created the video game "Pong" in 1972, and four years later Steven Jobs and Stephen Wozniak assembled the first "Apple" home computer. These inventions stimulated business fortunes and business disasters, but, most important, they plugged our imaginations into a startling new vision of the future.

The history of the video-game business is filled with lessons for anyone who contemplates the future of chip-based businesses. Indeed, the first lesson could have been learned from the parallel history of calculators, which were also based on a simple microprocessor chip. Calculators originally sold for close to $200, but since the chip itself cost only a few dollars and the calculator was easy to design and manufacture, competition developed almost instantaneously. The end result—that prices and margins plunged dramatically, driving many companies to huge losses and many to go out of business—should have been a lesson to the video game manufacturers.

The second lesson could have been learned from the parallel history of the digital watch, which brought chip companies into direct competition with established businesses. The same story of rapidly plunging prices and margins plagued the digital-watch business, but it was here that the chip makers and wizards of high tech should have learned their marketing lessons. Marketing—the art or science of product development, pricing, promotion, and distribution—is precisely what Timex, Bulova, and other traditional watchmakers knew well, whether selling through jewelry stores or drugstores, and what microprocessor manufacturers did not know. In head-to-head competition with traditional watchmakers, the fast trackers from Silicon Valley mostly went bankrupt. Thus, the second lesson was that technology must be governed by marketing. Furthermore, most of the traditional watch manufacturers now include digital watches, but they do not dominate the lines, which tells you immediately that a great percentage of the population isn't the least bit interested in digital readout or articulation.

All of this was going on while Nolan Bushnell was developing a simplified version of the games played after hours on mainframe computers by campus and company computer cultists and hackers. Adapting a term used in the Japanese game of Go, Bushnell founded a company called Atari. In doing so, he confirmed the third basic lesson of electronic prediction, which is that we should never underestimate the powers of invention nor the entrepreneurial spirit. "Pong" was a new idea and became an instant hit in homes and arcades. Major retailers such as Sears, Roebuck were distributing Atari products by 1974, and by 1976 the company was turning over $40 million. By this time the parallels to the digital watch and calculator businesses were both apparent and inevitable, as if the businesses were programmed by a microprocessor of its own.

Everyone entered the game, from pinball manufacturer Bally, to toy manufacturer Mattel, to the purchase of Atari by Warner Communications, the large entertainment and communications conglomerate. In 1979 less than a half-million game players were sold. In 1980 over 4 million units were sold and by 1982 annual sales reached nearly 7 million, totaling a 16 percent penetration of American households. Nearly 60 million game cartridges were purchased by these nearly 14 million households, totaling sales of $1.5 billion. But already it was becoming very difficult to make money in the video-game business.

The fourth lesson was learned in 1983. Player sales dropped by over two million units, and dollar volume was cut in half due to falling prices and margins. Cartridge sales increased 30 percent but price wars kept the dollar volume flat. These results could have been anticipated at some point based on experience with watches and calculators, and prudent business instincts would have planned on the management of such an inevitable future, so that even if they misjudged the particular quarter of a particular year, they would have been reasonably prepared for it. But they weren't, and understanding why they were not holds an important lesson for future thinkers: One must never lose sight of the true nature, the *Ding an sich,* at the core of technological products and their attraction.

Atari lost over $530 million in 1983, and toymaker Mattel lost over $200 million, not only because of competitive pressures on prices and margins and because of imprudent management, but also because they grossly overestimated the nature and attractiveness of their product. From the beginning video games were not the poor man's home computer but the poor man's VCR, purchased because the VCR home entertainment system was too expensive. Video games were never a sophisticated toy for adults as much as men's magazines promoted them. They did excite teenagers, but not as a toy to be used at home. Playing video games away from home in arcades and other teenage hangouts became the rage among some young people. After the first enthusiasm, however, the average household equipped with video games spent only twenty minutes a week with these entertainments—games that were becoming more and more repetitious, parodying one another simply because they quickly reached the limits of what you could do with a video game. But this kind of *Ding an sich* thinking never penetrated the Hollywood promoters who dominated the business—you had to spend big to profit big like DeMille. Against all evidence, they thought they were in something like the record business, and ended up with 35 million game cartridges they couldn't sell.

The industry is now regrouping, and it is too soon to count it out. But the bulk of the business will probably move to home computers, some of which are priced where game units were originally priced. But as we think about the future in terms of human behavior rather than corporate profits, we may already have discovered the limits of electronics as sheer entertainment—a steady 20 percent of U.S. homes.

The Home Computer

It is important to keep all these lessons in mind when imagining the home computer and considering the extravagant projections made about its impact on our domestic life. We must remember what a computer is in and of itself. The history of computers can be traced from the development of the abacus almost five thousand years ago to the invention of the Apple computer; it is a saga filled with men like the seventeenth-century philosopher-mathematicians Blaise Pascal, who created the first adding machine, and Gottfried Wilhelm von Leibniz, who invented an early calculator, discovering new mechanical relationships between numbers and spaces and spatial expressions of algebra.

ENIAC, the first all-electronic computer, was built in 1946, contained 18,000 vacuum tubes, weighed thirty tons, and cost $10 million. The silicon chip microprocessor gave large computers more and more efficient computing power and made the smaller personal computer possible, for in less than the space of a postage stamp hundreds of the required on/off electrical processes could be performed in a logic first dimly perceived in the abacus. Through a combination of remembering data and a logical sequence of yes/no or on/off decisions, the computer can approximate human thought.

The key question, of course, is: What kind of human thought? The answer—in its simplest terms and with great respect for intellects greater than my own—is: Memory and applied mathematics. In this context, it is helpful to recall that these types of machines created and solved military codes during World War II by continuously storing and matching random words until some translatable pattern of meaning emerged. Many of the codebreakers were not mathematicians but classics scholars with prodigious memories and quick, tireless, gymnastic minds.

To this human talent to store, retrieve, and manipulate data comparatively, the computer also adds calculation in a manner not dissimilar to looking up multiplication tables in the back of a book (or in the back of our memories). It does this at electric speed, and the machinery can even deal with words by giving a certain binary code to the identification of the letters of the alphabet. The computer can expand on these human talents for abstract memory, comparison, and calculation beyond mortal capacity or endurance. What it cannot do is change what the basic human talent is.

Jobs and Wozniak's achievement was to incorporate a strikingly large amount of capacity into a computer that cost the same or less than an automobile. Since it was individually affordable, it opened the door to the age of the personal, rather than the corporate, computer. Jobs and Wozniak also challenged the imagination to consider a computer as ubiquitous as the automobile—with an equally dramatic impact on society. They incorporated Apple Computer in 1977, and by 1980 annual sales reached $100 million. Late in 1980, they went public in one of the most successful stock offerings ever. The move not only netted Jobs and Wozniak nearly $250 million, but was also dramatic evidence of widespread investor confidence in a new kind of future, making 1980 the computer equivalent in many minds to television in 1950. The next ten or fifteen years would change the world.

The personal-computer industry didn't explode until 1982. Sales had gone from 150,000 in 1979 to 325,000 in 1980 and 500,000 in 1981. But in 1982 sales jumped to two million for a total of almost three million, or a household penetration of over 3.5 percent. In the process, of course, following the by now inevitable history of all chip-based businesses, a great deal of competition developed, with Apple and Atari joined by Commodore, Radio Shack, Coleco, Timex, and Texas Instruments, as well as IBM, Digital Equipment, and others. Prices were kicked around like a football, and customer confusion became rampant. Shopping for a computer was even more complicated than shopping for a car. Out of all this, however, something of a user pattern and a market pattern began to emerge, dividing into small business and home applications.

Compared to a fairly large corporate computer, the personal computer is a microcomputer with simple programs limited to doing one thing at a time, minuscule in storage capacity, and far too slow. But to the small-businessperson, these microcomputers were a dream come true, because they reduced what had been macro problems into micro problems. Record keeping, payroll accounting, inventory control, billing, secretarial functions, investment portfolios, and a host of other problems that plagued restaurateurs and doctors, lawyers and dry cleaners, architects and real estate brokers, suddenly became manageable, and in information-based industries such as finance, the small house could begin to compete with household names.

Computer manufacturers would have us believe that the benefits of their technology are improved efficiency, productivity, and planning. Perhaps this is the best way to describe the value of computer hardware to a large company. The individual worker, small businessman, or householder, however, is more likely to see the benefits of computers in the tasks they can perform. This approach shifts the focus from hardware to software. From this software perspective, the benefits most often listed for personal computers are payroll systems, accounting, business planning spreadsheets, data filing and manipulation, inventory control, tracking of stocks and bonds, tapping into Wall Street information banks, and word processing.

But few of these benefits—jobs that can be done better with a personal computer—apply to the home. Home uses are different from those of small or private businesses and home users represent a different market. As a result, most personal computer manufacturers have found themselves floundering as they try to present the benefits that computers can offer to home users. The manufacturer's ads focus on the entertainment of computer games and the instruction of learning programs. Even mighty IBM finds itself advertising little more to home owners and parents than that they should buy a personal computer so that they can enjoy themselves while giving their children a leg up on their education.

In survey after survey in homes across the United States, 50 percent to 60 percent of the people reported using their personal computers to play video games, and 30 percent to 40 percent employed them as an adult or child's learning tool. The next most important use, running close to second, was for balancing the checkbook or for family budgeting. Way behind were the 10 percent of uses reported as storing recipes, keeping an appointment calendar, or calorie counting. This is the extent of strictly home uses both people and the industry have found for computers to date. The same surveys report about 45 percent using their computers for office homework, while another 25 percent use them for home-based business applications, and a little less than 20 percent in word processing applications, followed by things like keeping mailing lists and information storage and retrieval. But when you subtract these almost exclusively business- and professionally oriented uses and concentrate on the home, it is not surprising that over 50 percent of all software packages sold are entertainment programs and another 15 percent are educational programs.

If home computers are used half the time or more for video games, we already know something of the upward limits of the market. If we add the calculator and typewriter uses for homework, banking, and budget balancing, we not only get a further picture of both upward limit and price sensitivity, but also how little difference it might all make in the near future. Nevertheless, personal computer sales doubled in 1983 to almost 5 million units. Oversupply and plummeting prices led to losses of hundreds of millions of dollars, and Texas Instruments, in one example of history repeating itself, withdrew entirely from the low end of the market, effectively dumping hundreds of thousands of machines. Sales fell to just below 5 million in 1984.

By the spring of 1985, IBM had ceased production of its PCJr and Steven Jobs had lost operating control of Apple Computer, Inc. The new president of Apple, John Sculley, declared that "a lot of myths about computers were exposed in 1984. One of them is that there is such a thing as the home computer market. It doesn't exist. People use computers in the home, of course, but for education and running a small business. These are not uses in the home itself." In 1985 such former heresy was becoming orthodoxy.

Looking to the future from the vantage point of past sales and experience, one can expect eighteen-month cycles of stable to plummeting prices as competitive products are developed and come to rest in a kind of parity. Much the same will happen in the software arena, but we can look forward to striking innovations such as programmed fiction that might satisfy the same instincts as Robert Ludlum's work does, and reference works, as well as video games produced in conjunction with the videodisc. All in all, it is reasonable to expect personal computers to have found their way into about 20 percent of our homes by the turn of the decade, which is not an ungenerous number if we discount for some not-so-small business purchases and applications. There is an upward limit, however, and it would be imprudent to project personal computers as being in more than 45 percent to 50 percent of our homes by the year 2000, especially since only about 25 percent of consumers under thirty have expressed any interest in having their own free-standing computer.

Michael Crichton, who enthusiastically calls the home computer "the best toy in human history," also warns that "computers are really a rather trivial part of human experience." This may be true, especially when we look at what a freestanding computer really is in and of itself. But when the home computer joins the other new electronic, personal-use media in significant numbers in our homes, they might very well change the way we think and act.

6

Dialogic Media

The new electronic medium that has the potential to transform our homes and offices into a totally wired and interconnected computer society is videotext. Of all the new electronic media, it is the least understood and the most difficult to project into the future. Its possibilities seem limitless, but it is difficult to predict what shape it will take in the American home since its success is highly dependent on how users are willing to change their habits of behavior. Nevertheless, many predict that videotext is the key to the electronic home of the future.

What Is Videotext?

Videotext is the generic term for the connection of the home television to a distant computer so that the user can either retrieve computer-generated information on demand, or both send and receive information in an interactive dialogue with the computer. One imaginative scenario shows the power of the computer reaching into the home to transform television sets from entertainment appliances into information appliances. Another view turns these "electronic cottages" into a "wired nation" with every home and business interconnected in such a way that we will never have to leave our homes.

Neither view entirely includes or excludes the other, but both assume a convergence or synthesis of computing and communications technologies into a new hybrid technology of household service. It isn't the individual technologies but how we assemble and package them that will radically affect our thought and behavior, reaching into the consumer marketplace, our home and family life, as well as into our workplace and business life.

There seems to be little question that if videotext happens it will change the conditions of our lives. There also seems to be little question that it will happen. It is frightening yet entertaining to speculate and dream about the possibilities. We have already seen how data processing has been changing our office lives and what robotics has done in factory production. We recognize that our bank statements and electric bills are printed by computers, that our credit cards are processed by computers, and that in a growing number of locations we both operate and receive real green cash from automated teller machines. We know that computers talk to one another between office and office, company and company, data resources and end users. Some of us know that newspapers are automated from the reporter's terminal to the printing press. We are aware of the growing market for personal computers, that telephones are universal, and that cable communications are on the verge of ubiquity.

There seems to be no reason why these growing resources cannot be synthesized, why we cannot simply plug one into the other and then plug our homes into the mix. In fact, it is very easy to imagine the computerized home of the future in a computerized society, but it is very difficult to imagine how it will function, how we will adapt to it, and how we'll get there. Even after millions of dollars have been spent in elaborate test applications, no one knows what the products, services, or markets will actually be for videotext.

Our purpose here is to grasp the new communications technologies in such a way that we can predict the pace of their arrival, the ways they will interact with each other, and their impact on our culture and personal lives. No technology demands more specific analysis than videotext. No technology demands greater concentration to comprehend its *Ding an sich*. No technology places such demands on our imaginations. We must each boldly project the possible courses videotext might take into the future while we gently and patiently restrain our wilder fantasies. It is an exercise in imagination that can tax both mind and heart. The technologies involved are multiple and complex, and because at root they offer a two-way system of communication, our normal one-way analysis of who will be saying what to whom may be inappropriate.

For example, when we analyze the process of communication between author and reader, playwright and audience, politician and voter, we have a clear one-way "communication" going from speaker/writer to listener/reader. Videotext as a distinct thing in itself is a medium of two-way communication, allowing for dialogue, conversation, exchange, the give and take of Socratic inquiry. For this reason, it is best to think of all the technologies and services grouped under the heading of videotext as *dialogic media*. Paralleling the postal service and telephone, they create an elaborate infrastructure for communicating with one another.

Before we begin to examine the various dialogic media, an understanding of the relevant vocabulary is essential. Vocabulary is tricky and important when talking about these media. *Videotext* with a *t* is the larger generic term that embraces both teletext and videotex. *Teletext* is one-way. *Videotex* without a *t* is the term used for two-way systems.

The Origins of Teletext

Television pictures are composed of a series of dots and lines built up so fast and continuously that the viewer, unaware of the still frames clicking past, experiences motion. When the horizontal control goes out of balance, we actually see a rolling series of pictures with black space between them. This space between the picture transmissions is called the "vertical blanking interval" and in the 1960s British engineers found a way to fill this otherwise useless space with data.

The BBC launched a system called "Ceefax" in 1976, and the independent broadcasters in England developed a parallel system called "Oracle." Both services required a modest conversion of the television set, which in a country where most sets are rented was viewed as relatively practical and efficient. Both systems worked essentially the same way. Quite simply, they aimed and fired little bursts of data—words, numbers, graphics—at the vertical blanking intervals, careful not to hit or interfere with the lines creating the continuous flow of the soccer matches, the news reports, or the weather forecasts the viewer may have been watching. However, if you wanted to know the scores of other soccer matches going on at the same time, for example, you pressed a few buttons on a keypad the size of a hand-held calculator and the sports data being transmitted in the vertical blanking interval would fill the screen.

As interest in these technologies developed around the world, they acquired a generic name. Since they provide textual data through a television broadcast, they were called *teletext*.

At the time teletext was being developed in England, the American imagination was preoccupied with UHF, pay television, and cable. Consequently teletext appeared both unnecessary and uninteresting. The availability of alphanumeric channels from the Associated Press, Dow Jones, United Press International, and Reuters, or community bulletin boards and public access channels on cable television, even made teletext seem useless. Furthermore, the time and effort required to modify a sufficient number of home television sets and the lack of urgency to do so made the effort too expensive to contemplate.

No one doubted that if a national or local teletext system was put in place, and if consumers used it to any great extent, advertisers would be willing to finance it. After all, the answer to the prototypical American question of "Who will pay for it?"—*it* being broadcast programming in a sender's medium—is "Advertisers." What lay in doubt was consumer need or interest in what teletext had to offer and the advertiser's willingness to capitalize the venture in advance and sustain it through a long gestation. Consequently broadcasters did little more than assign their engineers to keep abreast of teletext technology and appear in Washington every now and then to protect their regulatory flanks.

Two things were to change all this in the mid-1970s and early '80s: a public-interest lobby for the hearing-impaired and the development of wire-based two-way home data systems. Both generated an excitement that accelerated the development of teletext.

There are 14 million hearing-impaired and almost 2 million deaf Americans who had effectively been shut out of the television experience, unable to laugh at the jokes and cry at the tragedies of television drama. Individuals who work for and with the hearing-impaired joined forces with broadcasters to solve the problem, and basic teletext technology was the answer. Attempts to utilize simultaneous sign-language translations, as in some Presidential addresses and religious broadcasts, proved to be impractical on a continuing basis or for regularly scheduled commercial programs. But subtitling, as in a foreign movie, was practical from an aesthetic point of view, and if it could be done without disrupting the broadcast image in other households, it would clearly be a service in the public interest. Supported by foundation grants to public broadcasters and focusing first on translating the news, the industry perfected what came to be called "closed captioning" (CC) in 1979.

The commercial networks eventually joined in, and today the hearing-impaired can enjoy a handful of television programs each evening through the application of teletext.

The development of closed captioning derived from the perception in America that there was social value in providing access to television programming for the hearing-impaired. People with different political and technological expertise joined forces in providing a new concept in communications for a small but sizable segment of the population. Their efforts also put into place a foundation of broad technical and operational familiarity for the development of commercial teletext systems. More elaborate experimentation and subsequent announcements of national services, however, probably came from the competitive pressure from a growing enthusiasm for videotex.

The Origins of Videotex

Videotex also began in England, developed by engineer Sam Fredida at the research center of the British post office in Ipswich. It was launched in the late seventies under the name "Viewdata," which still may be the best way of describing a system intended to provide data from a remote computer for home viewing. The genius of the system is its at-home simplicity and relative low cost. Computers had long been linked by telephone, microwave, and satellite for business communications and data processing, but the method of linking the computers was expensive and the user had to be fairly well trained to enter the sequence of commands that are required to access the information.

What Fredida developed was a simple and inexpensive method of connection, low-cost terminals as simple as a push-button phone, and a menu-driven means of accessing computer information. In the jargon of the trade, it was "user friendly," for as soon as the terminal and television are turned on, the system tells you what you must do next (by pressing a single-digit number on your keypad). The program is built in a tree structure that leads you through a series of index pages to the precise information you are seeking or the transaction you wish to make. From the moment of its invention, it was perceived as the most friendly face a remote computer could have, and therefore as the inexpensive key for creating a market of interconnected consumer households. Viewdata provided a solution to the problem of developing a simple system for the lay person at home. The motivation for developing such a system, however, did not come from a marketing analysis of consumer wants and needs, but rather from the British post office's unique profit-and-loss situation.

Since the British post office owns and operates the British telephone system (and, in fact, has changed its name to British Telecommunications to reflect this), and since almost everywhere in the world the post office loses money and the telephone company or division makes money, it was perceived to be to the British post office's financial advantage if "occupant" or junk mail could be taken out of the mail carrier's pouch and put on telephone lines. This would also hold true for the bills and checks that constitute so much of first-class mail, as well as direct-response fliers, circulars, and catalogues. Postal administrators rightly concluded that banks and others would look upon such in-home electronic transactions as cost effective in terms of backroom bookkeeping and reducing front-desk, consumer-contact personnel. It all hinged, of course, on the government bearing the cost of establishing and launching the system.

A public Viewdata trial began in 1976. When the post office instituted full commercial service in 1979, they introduced it under the trade name "Prestel." Viewdata developed in France, Canada, and Japan, and today those countries, along with Denmark, Finland, Hong Kong, Holland, Spain, Switzerland, Venezuela, and West Germany, have all launched government-sponsored or -encouraged videotex systems.

Not surprisingly, the first American experimentation with

videotex was conducted by AT&T. It ran for six months, from August 1979 through the following February. Set in Albany, New York, and called Electronic Information Services, this experiment provided computer-terminal access to telephone-company information services such as white- and yellow-page directories and public-service data bases. The announced purpose was to test public reaction, but the true purpose was to gain technical feasibility experience from a consumer-use test. It was a technical experiment rather than a market test. The possibly unexpected result, however, was to alert, alarm, and arouse others throughout the publishing industry to what they perceived as a monopolistic thrust into their own businesses, particularly classified advertising. A year later, a similar AT&T experiment proposed for Austin, Texas, led directly to the newspaper publishers' successful litigation against the perceived encroachment.

In contrast, the program sponsored by the French government to equip all telephone subscribers in the Ille-et-Vilaine region of Paris with a black-and-white directory-assistance terminal as a substitute for paper directories has met no resistance. It may not prove so easy in America. In Albany and Austin we had the first evidence that competitive companies will not easily grant telephone companies such a commanding position in the American home of the future. In fact, they will use all of their legal, legislative, and regulatory influence to prevent it. Ultimately, all of the interested parties will compromise, for they need each other. But when this will happen, or how such a mass market will be structured, is far from evident. The debate will go on.

The only involvement in videotex by the American government was a one-year experiment that began the spring of 1980 when the U.S. Department of Agriculture, along with the National Weather Service and the University of Kentucky, launched its Green Thumb project. The project provided weather and crop information, commodities prices, and county extension services to two hundred farmers in Todd and Shelby counties, Kentucky. The experiment captured the American imagination, for it addressed a realistic consumer need, articulated the facilitation of remote transactions, and targeted its marketing to a definable population segment that already had a disproportionate number of home computers, for in fact, these farmers were small-businessmen. The main value of this experiment was that nearly every interested party could support it and talk about it without prejudice precisely because no one believed that such a public-interest-driven and government-financed system would ever become a general reality in the United States.

Types of Competitive Mergers

A few months later the first commercially oriented experiment was launched in Coral Gables, Florida. The joint project sparked a new sense of urgency, for although everyone knew that the consumer market was years away, there was renewed concern over whose market it would be and what competitive moves must be taken now to ensure one's choices in the future. Each of the major tests of videotex over the next three years would represent a different combination of types of players, technical configurations, and marketing assumptions. This first exercise joined Knight-Ridder, by some counts the largest American newspaper chain, with thirty-five newspapers, including Florida's *Miami Herald,* with AT&T. Knight-Ridder played the major role of collecting and editing the data, while owning and programming the computers for which the telephone company simply provided the terminals and interconnection.

Called "Viewtron," the Knight-Ridder test ran from July 1980 to September 1981, among 204 households with 674 people who accessed 1.5 million pages in 25,500 distinct sessions on the system. This meant that, on average, a household accessed the system about twice each week and viewed nearly sixty pages on each occasion. Twenty-five percent of all the pages accessed were news, reported by consumers as the most important category. Next in order of preference was the community bulletin board, on which people also placed their own messages, and the guides to local entertainment, events, food, and dining. Shopping ranked seventh. Although 83 percent reported that videotex was a convenient shopping medium, only 68 percent of the trial households actually purchased something through the system. Nearly a thousand orders were placed and the average expenditure per household was $62. Overall, 90 percent of the participants said they liked the service and that videotex improved their ability to obtain information and services.

Almost all the services people used provided information usually found in such printed materials as newspapers and telephone directories. These results support one of the key assertions of the Viewtron research, which reported that 90 percent of the system's users said they used other media less when videotex service was in their home. Television viewing eroded in 45 percent of the households, but, more important, 33 percent used newspapers less, with catalogues, yellow pages, and magazines following. Similar research by others has led many print publishers to believe that videotex will be cannibalistic, devouring anywhere from 30 percent to 60 percent of their current revenue. This belief alone will lead to a great deal of regulatory lobbying, litigation, and investment.

Times Mirror, a company with annual revenues in excess of $2 billion derived from major newspapers such as *The Los Angeles Times,* from book publishing, data-based information services, and extensive cable television holdings, represents a second kind of company vitally interested in videotex. This major print publisher also owns its own cable wires reaching hundreds of thousands of consumers. During nine months in 1982, Times Mirror tested a videotex system called "Gateway," in 350 homes in Mission Viejo and Palos Verdes, California. They offered the same package of banking, news, shopping, games, bulletin boards, and electronic mail on a 70,000-page data base. What made this test significantly different was that it used cable television as its wired connection, although in some instances it also used AT&T lines.

Subscribers are reported to have used the system several times each week, with an average of twenty hours each month. The services subscribers liked and used the most were games, electronic mail, news, shopping and banking, the information guides to entertainment, and electronic mailbox addresses. These services represented in excess of 70 percent of the total usage. Most interesting was the consumer dynamic. Games had distinctly highest usage at the beginning of the nine months, but were replaced by news as the leading use during the Falkland Islands crisis and the 1982 elections. By the end of the test, however, electronic mail was the primary use. The vast majority (87 percent) said they felt quite secure in using the banking services, and 53 percent purchased something through the system, but on average they spent less than $100 during the nine months. Most popular purchases were for theater and airline tickets, books, and videocassettes.

A third possible combination of competitors was introduced when CBS, the most successful broadcast network with over $4 billion in revenues, joined forces with AT&T to test videotex in Ridgewood, New Jersey, beginning in September 1982. In this instance, CBS was responsible for the editorial content but AT&T controlled the computerization as well as the terminals and distribution wires.

Variously called "Venture One" and "Reach," the CBS/AT&T experiment extended from September 1982 through April 1983, among 200 households. Reportedly, nearly 2 million frames were accessed in over 47,000 sessions with a little more than half of the households active each day, accessing an average of 166 pages in one hour and twelve minutes. Between 20 percent and 25 percent of the adults in the test accessed the system on any given day, and more often than not, family members used the system together after dinner. By far the most popular feature was electronic messaging, followed by shopping, games, and weather. News was viewed by less than 25 percent of the households each day. Banking was perceived as a popular feature, but consumers gave the service a negative ranking, for it did not provide a hard-copy record of their transactions. During the second phase of the experiment, 42 percent purchased something through the system. In short, allowing for variations in the content and promotion of the services in the Knight-Ridder, Times Mirror, and CBS tests with and without AT&T, the results of all the experiments to date are generally compatible and provide similar clues to consumer interest and behavior.

A fourth configuration and a different technological bet was made by Time, Inc., unquestionably the largest force in cable television at the time as the owner of the second-largest MSO (Multiple Systems Operator), American Television and Communications, as well as Home Box Office, Cinemax, and a one-third interest in the USA Network. Time concluded that videotext was both unnecessary and too expensive and initiated a test of an enhanced teletext service through its cable systems in Orlando, Florida, and San Diego, California. Enhanced teletext, providing for the one-way retrieval of information in the home, utilized a full cable television channel with a capacity of five thousand pages, as opposed to several hundred in a broadcast mode. After almost a year of drum beating and operational testing, Time canceled the project and withdrew from the competitive field for the time being.

Publicly Time argued that it was forced to abandon the project because of its Japanese supplier's inability to manufacture a terminal at a cost below $150, which Time believed essential. Anyone schooled in both new and old electronics knows that the ruse of blaming the terminal manufacturer is as common and suspicious as telling a bill collector that "it's in the mail." Time expected to charge the consumer about $10 per month and to leave this revenue with the cable system. Time Teletext was to obtain all of its revenue from national advertisers. Many concluded that the product and business concept were misguided and destined to be stillborn. Some research confirmed that what consumers really wanted from the Time experiment was videotex and two-way capability, but one mismanaged failure does not spell the end of a technology or eliminate its possibilities.

One of the obvious results of all of the videotex experiments and aggressively promoted business plans was to encourage parallel teletext experiments. In 1983, both CBS and NBC announced national teletext services, apparently satisfied that the public was ready for instant news, sports, weather, business information, and comparative store prices. Other popular pages were entertainment listings, movie reviews, and personality features. The networks provide national material and sales, and the local broadcast affiliates supplement local information and advertising sales. Both commercial networks are currently providing national teletext services, but at this point few of their affiliated stations are equipped to provide the service locally. For the moment, neither is pushing consumer terminals and both appear to be waiting for Zenith and RCA to provide new television sets preequipped with the necessary decoders.

Videotex is a much more complicated and capital-intensive business than teletext. Not only is the technology more elaborate, but the hundreds of thousands of pages involved require the coordination of multiple information providers. Who the dominant coordinators will be, who will provide the interconnecting hardware, and whether the emerging competitors will provide compatible systems are still seriously in question. The basic questions revolve around: "Whose wire will it be?" and "Who will bundle the information and transaction services?"

Imagining a Viable Delivery System

Some believe that cable television will be the preferred wire because of its electronic capacity, its growing ubiquity, and the fact that utilization of its wires does not tie up the telephone. Unfortunately, as cable grows, it will limit the growth of broadcast teletext, for at present cable systems are not required to carry signals in the vertical blanking interval and, in fact, most cable systems have allocated this space for other purposes. Those who are cable subscribers are both demographically and psychographically the most likely customers for teletext. Thus the more cable grows, the more it may shrink the potential teletext market, at least in the near future.

A further consideration is that cable television is not presently a two-way medium. By 1990, however, something like 80 percent of the systems will be two-way *capable*. But it costs about $1,000 per mile to upgrade a capable system into a two-way active system. At the normal rate of 50 or 60 subscribers per mile of cable, that means it will cost the cable operator about $1 million in construction to provide 50,000 subscribers with two-way television. All of the videotex research to date indicates insufficient consumer demand to encourage most cable operators today to undertake that level of capital investment.

The telephone is both ubiquitous and two-way, and it is easy to imagine it as the foundation for interconnecting us in the future much as it interconnects us in the present. The first synthesis of telephone, television, and computer occurred in the early 1960s when AT&T began transforming itself into a digital network. The distinction between processing information in a computer and communicating the same information disappeared because the communications have been reduced to the same series of bits and bytes used in computers. Today graphics, sound, and full video pictures are transmitted in the same manner as the original letters and numbers. The electronic pulses that send, receive, create, and store information look exactly alike as they pulse along the wires. From this technology AT&T developed "packet switching," enabling senders to distribute information along the least trafficked path of the moment so as to increase the capacity of the network and reduce costs.

The dream, of course, is that this wired network of telecommunications that has served industry so well will become the webbing of home computer networks. But there are certain problems. The first is the degree to which the existing telephone lines could handle the expected increase in load if the dream of a nation of electronic cottages comes true. The second is the question of whether or not consumers will be willing to tie up their telephones in order to use their computers, especially when confronted with telephone time costs.

It is conceivable that consumers will either pay for two telephone lines into their home, dedicating one to their computer as some parents now do for teenage children, or take advantage of direct broadcast satellites. But most dreams of home telecommunications networks are premised on the replacement of the telephone company's copper wires and the CATV companies' coaxial cables with optical fibers that are composed of glass threads no thicker than a human hair and that use bursts of laser light to distribute digital information.

One such thread can carry ten thousand times the information carried by one pair of copper wires or 200 television channels, with room left over for two-way communications. But the prospect of significant transformation of copper to glass fiber is limited by construction time, cost, and the real balancing of need and demand with cost. Irving Kahn, who was the czar of cable television for more than a decade as the founder and president of Teleprompter and who is now a leading proponent and manufacturer of fiber optics, candidly estimates that it will be at least twenty years before 50 percent of Americans have fiber optics available in their homes. As Kahn sees it, "The issue is two technologies: the technology of technology, and the technology of politics. The second will be far more important." It is difficult to predict the outcome of the inevitable lobbying, litigation, and regulation, but without a doubt it will impede technological growth and create confusion in the consumer marketplace.

In addition to all these videotex and teletex experiments,

there were several actual home-based personal computer data and transactional businesses conducted through telephone and cable lines. Leading among them were CompuServe, now owned by H&R Block, Dow Jones financial services, The Source, now owned by Reader's Digest, and Comp-U-Card's Comp-U-Star. At one end of the spectrum, offering pure information, is the Dow Jones combination of *Wall Street Journal* and *Barron's* data costing $22 per month and accessed through cable. The service has about 60,000 subscribers and adds about 1,500 subscribers each month. At the shopping and transactional end of the spectrum, Compu-U-Star has about 50,000 subscribers who can access merchandise listings of over 30,000 products from about 200 manufacturers. These products costing 10 percent to 40 percent below the manufacturer's suggested list price can be ordered through the system or paid for by check or credit card. The computer actually shops for you as it searches its data base for products to fit your specifications. The service is available through the Tymnet network in 250 cities and any number of computer models can be used as the access terminal through the telephone lines. None of these "closed user groups," as they are called, however, has more than 100,000 subscribers today, and even at the total rate of growth of 10,000 to 20,000 subscribers per month, these systems do not seem capable alone of forming the foundation for a mass market.

One might wonder why, if all of these market experiences seem to point to a conservative picture of the future of videotex, the technology has built up such a head of steam. Part of the answer lies in the fact that after years of strategic research and development, many companies are driven into videotex services by their view of competition. If the videotex services that consumers favor and use the most are precisely the services newspapers now provide, major newspapers cannot afford to repeat the oft-told story of the railroads who saw no competitive threat in trucks and airplanes and failed to perceive themselves as being in the transportation business. Newspapers are in the information-service business and have no intention of allowing this business to be slowly eroded without their participation. At risk of losing their existing business, investments in new businesses based on new technologies may seem modest.

Newspapers already have an embedded investment in internal electronic systems and the largest now receive and print their news by satellite. Combined with the various telephone, cable, and computer companies that share the point of view that the best defense is aggressive investment, major coventures will emerge to subsidize the transition to consumer electronics. For example, to a company such as Times Mirror, both a major publisher and cable MSO, the idea of spending $20 million to create the wired foundation for a one-million-customer videotex service is neither outrageous nor impractical, though to a pure cable operator it may seem so. It will happen.

Videotex is emerging as a classic merchant push—rather than consumer pull—business. Banking is one of the clearest examples. Financial transactions comprise about half of the 100 billion pieces of mail handled each year. Banks have increasingly turned to electronics, and in the striking growth of automated teller machines, banks have established the technical capacity for home banking services. Bank of America, Chase Manhattan, Chemical Bank, Citibank, Manufacturers Hanover Trust, and Security Pacific have invested in videotex and closed-user computer tests and experiments. They have discovered that banking alone will not justify videotex financially, but when combined with financial news and other services, it becomes very attractive to the consumer. If a bank could simply eliminate the 85 percent of customer-service phone calls for routine information such as banking hours and locations or confirmation of check clearance, they might be willing to invest in a system for a considerable period of time.

The cost-avoidance strategies are further supported by bank deregulation, allowing the major banks to position themselves strategically for the nationwide consumer banking made possible by electronics. These competitive motives might well support the construction and maintenance of a national videotex service that operates in the red for a number of years. We can benefit from recalling that in 1904 Edward F. Hutton, out of sheer self-interest, advanced Western Union the $50,000 required to establish the first coast-to-coast wire. Hutton wanted the opportunity to open branches across the nation. Proportionately, the amount required today is not significantly greater, and banks and investment houses once again have the motivation to assist in wiring the country for the future.

The other driving force pushing the market is direct-response advertising. Direct mail is a $40 billion business and growing so rapidly that many project that by 1990 one-third of all merchandise sales will be out-of-store purchases. The opportunity to combine the merchandise catalogue display, the price offer, and the transactional response in one medium is enormously attractive, even if the sales results evidenced in the videotex experiments are not overwhelming.

Based on the assumption that these various forces driving the technology will make videotex available, projections of videotex penetration range from 8 million to 12 million households in 1990, and 30 million households generating a $30 billion business in the mid-1990s. All of these projections differ not only in their final numbers but, more important, in their assumptions about content, services, pricing, construction time, methods of distribution, terminals, standardization, and regulation.

CBS, for example, has concluded from its test that the results support "forecasts that put the market potential for videotex in the 20 million plus household range by the year 1990." Unfortunately, this sounds suspiciously like an estimate of the number of home computers in place, rather than an estimate based on consumer behavior and intent to buy data obtained from 200 homes in Ridgewood, New Jersey. Viewtron was launched commercially during the late summer of 1983. With 5,000 terminals for sale at $600, it was the first to ask consumers to pay for the service. A year later only 2,800 South Florida customers had subscribed. Since then, Viewtron has cut its staff, made its system compatible with various home computers, and cut its monthly subscription costs. With this arrangement, Viewtron is now available nationwide, with advertisers such as American Express, Buick, and J.C. Penney supporting as many as 150 data pages each.

Other partnerships and nationwide systems announced include AT&T, which has teamed up with Chemical Bank, Time, Inc., and Bank of America. Their system is called "Covidea." CBS has joined forces with IBM and Sears, Roebuck in a venture called "Trintex." A joint venture of RCA and Citicorp lost a partner in J.C. Penney, which has now joined forces with GTE. Targeted for 1987 launch, all of these ventures support the understanding of videotex as a manufacturer-push rather than a consumer-pull service.

Meanwhile, almost every major bank in America is offering computer banking services to home computer owners for $5 to $15 a month. Many believe that these customers will rapidly convert to videotex services once they are available. But the fact remains that we really do not know what the consumer response will be, or how willing both consumers and merchants will be to change their behavior in any major way. What we have at this point is a fairly clear map of where the pushing will come from, when it will come, and why.

Videotex has bedazzled many of us with its high I.Q. Not only are videotex services of all varieties inherently intelligent, they are intellectually glamorous, opening up apparently limitless possibilities and forcing us to rethink fundamental strategies. Not only do they threaten historic businesses based on older technologies, they trigger new public policies and regulations. Not only are the various services and technologies individually complex, they compete for the same investors and, eventually, consumers. Consequently it is not surprising that videotex has mesmerized journalists and their readers into speculating about unrealizable fantasies for the future.

If we consider all of the technologies and configurations that make up the dialogic media as one, it is reasonable to project that they will be serving 10 percent of our households by 1990. This level of penetration will result from the force and weight of competitive investments made in jockeying for position. After 1990 the market will take on some shape, and competition will have resolved itself along clearly defined and regulated lines. A business base will have been established as well as new businesses. At that point, nothing should stand in the way of videotext growth, and penetration levels of 40 percent to 50 percent can be expected by the year 2000.

It is in that decade that the conditions of life will begin to change dramatically, for videotext has the power to radically transform the very conditions of our lives, our value systems, even how we think. With these technologies we will be able to engage in basic activities such as remote information retrieval, messaging, business and financial transactions, and computing. We will enjoy telemonitoring not only for burglar and fire alarms, but for the regulation of home appliances and utilities. A new dialogic and electronic way of life will emerge that will truly challenge both our character and personality.

7

The Video Environment

Television was conceived in the phosphorescent glow of display sets in appliance-store windows and the small audiences standing outside them on summer evenings. The early programs were not as compelling as the technology itself. When the first annual Emmy Awards were given out in 1948, Mike Stokey's "Pantomime Quiz Time" was voted the Most Popular TV Program, and the statuette for Most Outstanding Television Personality went to Shirley Dinsdale and her puppet, Judy Splinters. Both Emmy winners appeared on independent station KTLA, Los Angeles, which also won an award for its overall performance. There were no national networks. President Truman would not inaugurate transcontinental television broadcasting until 1951.

Yet on those summer evenings, the potential audience dreamed. They dreamed about the Christmas they would put a television set under their family tree, and they dreamed about

146

the hours of enjoyment it would provide for the family down through the years. They could see themselves turning it on, tuning it in, and settling down for an evening's entertainment, just the way they saw themselves driving a new car. The more they dreamed, the more they decided they could afford it; and each dream gave shape to the actual future—to all the programs, products, and services that we have come to call television.

The ultimate concern of this book is to understand social character as it is shaped and defined by communications technology. To grasp our future social character we must somehow see all of our communications at once without losing sight of the distinctive character and contribution of each. Unfortunately, this is when reality often evades us, and our dreams become outlandish fantasies. We tend to imagine all the media doing everything all at once, or one medium doing it all, and in our imaginative leap we lose ourselves. We need to project cautiously so as not to lose the role that we as thinking, feeling human beings will play in that future. The challenge is to fill the lost imaginative space with real people who will choose what to see and hear, watch and read, think and vote and buy, and to avoid the farfetched speculation that leads to appealing science-fiction scenarios that miss the point of personality, personal living, and reality.

In describing the new electronic media, we have developed a working hypothesis of their futures based on the technological, economic, and marketing facts of their past and present. Yet each working hypothesis is an isolated snapshot without motion, based on trend lines describing a static point in the future. The individual trend lines do not reveal how the media will interact, compete, and coexist. In other words, these isolated snapshots fail to show how we will actually experience the media. It is for this reason that we should return to those summer evenings in front of appliance-store windows. There lies an understanding of how purchase decisions are made and of how our imagination of a product's use brings it to life, shapes its character, changes our behavior, and forces a metamorphosis in all other media. We are the people who will both create and live in the emerging video environment, and we must somehow find ourselves within the probable data, and in the process bring the data to life.

For many years, the actual number of television sets sold in America was very much the same as the actual number of automobiles sold in the same years. These living choices can tell us a great deal about how we adopt new products and integrate them into our home life. Looking into an appliance-store window and looking into an automobile dealer's showroom were similar experiences. Both cars and television were complex, highly engineered products that were easy to use. Both products were expensive yet available for trial experience, from a test drive around the block to watching wrestling in the corner tavern. The advantages and appeal of both products were immediately observable. We all talked about buying cars and televisions, and we talked with everybody because the products were easy to describe; and there were few Luddites or Cassandras to dampen our enthusiasms. Both were 100 percent compatible with our values and life-styles, focused on home, family, and conspicuous consumption. In a very short time, a family was not perceived as fully equipped without a car and a television set.

The marketing history of cars and television sets tells us that the diffusion and establishment of new products and technologies in society is dependent on several key factors. Is owning the products and technologies perceived as an economic and social advantage? Are they perceived as relatively easy to understand and use? To what extent are they subject to trial experience? Are their benefits easily observed and easily communicated to others? And, last, are they perceived as compatible with existing values, needs, and past experience without demanding extensive changes in our behavior and life-style?

None of these questions can be easily answered regarding the new electronic media.

Choosing the New Electronic Media

Faced with the cornucopia of new electronics, the consumer will conclude that each can be: a friend, providing dialogue and entertainment; a teacher, informing and advising; a facilitator, helping with the tasks of life. In short, each is a plug into the world of people, events, and activity. And every person will ask which plug or combination of plugs will best serve his own needs, interests, and life-style. Without being consciously aware that we are doing so, we must each personally apply the abstract checkpoints of what is known as the *diffusion/establishment theory* to every product. Using informed imaginations to apply these standards to the communication media as a whole, it is possible to gain an overall picture of the future.

For example, it is possible to understand in a very personal way how cable television followed all the abstract rules of diffusion and establishment. Cable television shared many of the key characteristics of both television and other successful new products. Speaking to Jack Valenti, who represented the Motion Picture Producers Association, Senator Barry Goldwater commented at a Senate committee hearing, "Where I come from, Mr. Valenti, we need cable to get television, and therefore, we're going to have it, Mr. Valenti, whether you like it or not." In much of Arizona to have cable is to have television, and without cable the advantages of television are unavailable. In that situation, cable and television can be perceived as the same. For close to twenty-five years, however, this was not the situation in much of America. It was not until the mid-1970s that a majority of people started coming to equate cable with normal and expected television service. The quantitative and qualitative advantages of cable were clear. They were easily demonstrated, communicated, and used to enhance an already accepted lifestyle. The younger, more affluent, and more educated consumers who leverage our culture and social character rapidly chose cable as their plug into the world of television news and entertainment. Given the strong American belief that more is better, it was not that hard to imagine nor to predict that 60 percent to 65 percent of our citizens would subscribe to cable television when it became available.

As deeply rooted in our society as cable television appears to have become, however, anywhere from 30 percent to 40 percent of American consumers will not be cable subscribers in 1990 even though cable wires will be in front of most of our homes. This creates more of a challenge to our imaginations: Why will some of us hold out and not subscribe?

A great number of people are quite content with the television they received when they purchased their first set. It continues to fulfill its promise. Many people still buy consoles to cover the mark left in the carpet in the room where their previous television furniture stood. Time will close this generation gap, as it will probably allay the fears of those who think cable television is too risqué to be in a home with children. But even if these gaps close or erode, many families in areas of reasonably good reception and broadcast service will not feel compelled to pay upward of $30 a month for cable. Some will simply not be able to afford it. Furthermore, many young, affluent, educated people will pass up the opportunity to make this investment or will cancel their subscriptions after sampling what cable programming has to offer. They have a choice, can afford the service, and yet will reject it.

When we were looking at television sets in appliance-store windows thirty-five years ago, we saw only one thing: a dramatic new product we could easily and even anxiously imagine in our homes. We could envision ourselves using and enjoying it. When we look into a video-store window today, however, like a child walking the aisles of Toys-R-Us before Christmas, we see many things. This simple fact changes the question from: When will I put a television set under my Christmas tree? to Which of the many video appliances will I put under my Christmas tree this year?

We cannot imagine the future diffusion and establishment of cable without considering that many people are purchasing personal-use media such as VCRs, video games, and videodiscs because they provide a personal control and convenience that cable cannot offer. It is difficult to predict which of the personal use media, if any, will eventually be the popular choice. Even the VCR, despite its present dominance, may not end up being the key personal home entertainment technology. In fact, all of the new electronic media are directly competitive with each other, and to imagine our purchase decisions in that environment is to begin to feel the dynamic of the future and the actual diffusion and establishment of the appliances.

Unlike television, some of the new electronic media are not available everywhere. You cannot subscribe to cable television, for example, if the franchise in your community has not been granted. Nor can you subscribe to thirty-six channels with several pay-television options if a comparatively ancient twelve-channel system in your community has not been rebuilt. And should such a system be rebuilt and upgraded, many people will choose not to buy into it because in the meantime, they made commitments to other new electronic appliances that fulfill their life-style needs. They may previously have purchased a videocassette recorder and see no special advantage in pay-television programming. This especially applies to the dialogic media that will not be broadly available until the next decade. When they do become available, they will find themselves in a competitive environment dramatically different from what we can extrapolate from the situation today.

The only way to grasp the potential future of all the new electronic media in a vibrant, vital, involved, living way is to project our own personal answers to the choices that will confront us, and imagine millions of personal dialogues and decisions on the part of other Americans. We should imagine the simultaneous availability of all the new electronic media and ask, Which of them would I put under my Christmas tree this year and how and why would my neighbor choose differently? Once our preferences are clear, we need to take into consideration the staggered availability of the new media. Which choices that we might make today will eliminate or dampen enthusiasm for future choices? If we can answer these questions with reasonable certainty and grasp each medium's *Ding an sich,* it is possible to get beyond the sterile statistics, the marketing hyperbole, the macho journalism, and opinionated self-assertion that cloud our thinking about the future with groundless fantasies. This is the only way to arrive at an athletic understanding of the emerging electronic landscape in which we will live and communicate.

The Premise of Mutual Success

Most of what we read and hear today about communications media is as aggressive and defensive as the worst political rhetoric. There have been decades of rancorous debates between broadcasters and cable operators, Hollywood producers and videocassette manufacturers, theater owners and pay-television distributors. Newspaper publishers fought telephone-company involvement in videotex under the premise that it threatened their business, while at the same time many publishers were investing in videotex to protect their businesses. As the competing parties projected the success of various media in a kind of electronic jungle where the law is kill or be killed, the impression left was that one medium must die for another to succeed.

To imagine the future as a Darwinian struggle among competing technologies is to misread the past and to obfuscate the present dynamic in today's market. In fact, history supports a different view of media development, a view on which we can project our future—the Premise of Mutual Success, which states that *all the new media will succeed and all the old or existing media will survive.* This premise reflects the facts of past history, in which new media have always proven to be incremental, and it focuses our imaginations on the dynamic changes occurring at the personal level where we make significant communications choices. Displacement theories of success and failure suggest a future in which we have not changed what we do, but only how we do it. On the other hand, we know that to imagine a computerized future is to imagine a very different society and social character. The mode, type, quantity, and quality of our communications will change, producing a more dynamic and revolutionary environment than a narrow Darwinian explanation would suggest.

An old Talmudic proverb says that if you want to be taller than another man, do not dig a ditch for him to stand in—stand on a chair. In a similar manner, the Premise of Mutual Success focuses on what each medium naturally does best, identifies its intrinsic characteristics, and celebrates its *Ding an sich*. The more competitive the environment, the more each medium will articulate its unique strengths and collectively energize our communications environment.

The best recent example of how the Premise of Mutual Success could have predicted the content and character of current and developing communications, as well as our personal behavior, is the growth of pay television and its impact on broadcast television. At the same time that broadcasters and theater owners were lobbying against pay television as a fundamental threat to their businesses, the cable television industry generally doubted that consumers would be willing to double the cost of a cable subscription in order to have access to commercial-free first-run movies. We now know that they were both wrong. Not all consumers, but a significant number of them, have, to the surprise and delight of cable operators, become pay-television subscribers; and movie theaters and broadcasters are still profitably in business.

The premise that there was a successful market for all of these competitors had to imply certain things. First, it implied that people were willing to pay to see Hollywood films in their homes without the hassle of actually going to the theater. Second, it implied that consumers would still seek the enjoyment of going out and seeing films on a big screen. The Premise of Mutual Success would have proposed that the competition was in-home versus out-of-home, and that it would probably extend the film product's paying audience and profits.

This same Premise of Mutual Success would have offered a different scenario for broadcasters. Here the competition would be head-to-head in the home, with both broadcasters and pay-television operators appealing to the at-home audience. Naturally, releasing theatrical films in an uncut and uncluttered version on pay television would dilute the audience for the same theatrical films released later on broadcast television in censored and commercially interrupted form. However, pay-television operators had to consider that for subscribers to pay as much as $20 per month to have new motion pictures available in their home on television would require a significant measure of commitment and involvement with fresh motion-picture product and special-event programming. The key issue for prognosticators should have been to estimate the time when there would be a sufficient number of pay subscribers to cover pay-television operators' cost and to reduce the broadcast ratings of motion-picture films.

The Premise of Mutual Success, however, also implies success for the broadcasters and would have projected competitive activity based on broadcasting's strengths and willingness to do battle. This programming would emphasize real-time events such as the Super Bowl, awards ceremonies, and beauty pageants designed and promoted as societal experiences. When ABC did not have any professional major-league contracts, it created "Wide World of Sports" and "Battle of the Network Stars." This example of tactical response suggests that broadcasters would cease to rely on theatrical films and invest in original film product, which pay households had already expressed extraordinary interest in. Offering made-for-television movies and creating special-interest events has already resulted in new, socially involving miniseries such as *Roots, Shogun,* and *Winds of War.* The Premise of Mutual Success would also have suggested that the pressure from uncensored theatrical films on pay television would produce sensational, popular, made-for-television films such as *Lace* and *Master of the Game,* which pay-television subscribers watch as avidly as everyone else. The Premise of Mutual Success would also have implied that pay-television companies would discover a viable structure for scheduling their films to provide the maximum opportunity for their subscribers to see the films and justify their expense. This flexibility, in turn, would allow pay subscribers to pick and choose among network broadcasts without eroding the value of their pay-television subscription.

All of these projections of consumer and network behavior could have been made based on a candid and unprejudiced application of the Premise of Mutual Success. They have in fact turned out to be true and they will continue to be true as pay television grows. Much of this view has now become the accepted wisdom, and the Premise of Mutual Success retains its predictive power. The future of communications will be dramatically complex and dynamic. To understand our future personal and social environment, we must try to grasp the content, variety, shape, context, competition, and vitality of the media, and to insert ourselves in a participatory way in the future. The Premise of Mutual Success has demonstrated itself to be the most rational, imaginative, and practical tool for predicting the future. To believe it is to apply it.

The Premise of Mutual Success and Cable Television

If the cable television networks are going to be successful, their success will be based on the genius of the past and the genius of the future. Some cable networks will adopt the time-tested strategies of independent broadcast station, relying on old movies and former network series scheduled as "counterprogramming"—for example, scheduling reruns of old situation comedies against the networks' daytime game shows. Other cable networks will refine the genius of the future and take advantage of cable's ability to move one or two steps down from the absolute demands of mass programming; they will focus instead on rather broad-scale service programming with a sharp editorial edge, counseling us on everything from health, beauty, and fashion to home repair and financial investments. When these two strategies are successfully implemented, the broadcast networks will have to make competitive adjustments of their own, not unlike the competitive moves they were forced to make in their prime-time scheduling in the presence of pay television. Thus the production style of today's daytime soap operas will evolve to look more like "Dallas," "Dynasty," and "Falcon Crest," while still exploring leading social problems. There will be more new productions and perhaps the gradual elimination of game shows, talk shows, and network reruns in daytime. Successful competition from "counterprogrammed" entertainment and from sharp-edged, utilitarian service editorial will require it. One could argue that a growing senior citizenry will continue to demand the game shows these people now support, but advertiser-supported broadcast networks will probably abandon these program formats to independent stations and cable networks in order to focus on attracting the younger and higher-spending baby boomers. In either case, however, the Premise of Mutual Success frames our participatory dialogue with the future and outlines the quantity, style, and content of future programming choices.

Current research also indicates that people defect in droves from the network reruns, at least from June to September. This happens especially in pay television households during prime time. As the cable networks become more vital, however, this will happen during the daytime hours as well. With the long rerun season, the networks risk a reduction of their ratings performance and income, while the cable networks, pay television, and independent stations find their audience and establish new viewing loyalties that could have a serious negative impact on the networks during the other three seasons of the year. Slowly but surely, this competitive situation will force the broadcasters to extend their new programming schedules into the summer, or to discover new programming options.

The fundamental reason for network reruns and for the erosion of the first-run season from thirty weeks to twenty-six weeks to barely twenty weeks today is economics. These cutbacks were made possible by unspoken agreement of the broadcast networks to play by the same noncompetitive rules. The broadcast networks claim that as the cost of original programming escalates, they must get a minimum of two uses out of program production to recoup their investment. However, as the economic benefits of reruns are outweighed by the economic threat of the long rerun season, broadcasters will look for other ways to save money. In the most recent round of lobbying for deregulation, the networks demanded that they should be allowed to share in these programs' syndicated revenues to help offset their original production investment, which was becoming impossible to recoup during its network broadcast exposure.

The Premise of Mutual Success recognizes that economics responds to market conditions; but it also recognizes that new technologies create new markets, and that new markets create new economics. The base fact is that the broadcast networks will be forced to develop vital new programming strategies to maintain their share of the audience. Several new markets and technologies will make this economically possible. The first is the economic value of syndication itself. The audience for independent stations has grown—thanks in part to cable television—and the prices paid for former network series in syndication, such as "Magnum, P.I." and "M.A.S.H.," have become astronomical by prior standards. Second, we are reaching the point where every broadcast station is equipped with a satellite earth station. Since a major part of the cost of syndication is in its stagecoachlike distribution, the substitution of efficient satellite distribution, while not necessarily reducing the overall cost of syndicated product, will increase the percentage that goes directly to the program producer's profit. Third, not only do cable networks offer a new level of distribution profit, but as the cable networks grow in size and audience, this tier of distribution becomes more valuable. All of these projections of the Premise of Mutual Success of independent stations, of syndication, of satellite distribution, and cable networks suggest that broadcast networks will be able to adjust their percentage of a program series' capital cost downward and liberate the funds necessary to initiate more original programming. It is also likely that, under the circumstances, the regulations covering a network's participation in syndication will be relaxed. In short, a complex and dynamic view of the Premise of Mutual Success for multiple distributors and technologies supports the projection of a radical reduction of rerun programming on the broadcast networks.

The combination of satellites and independent stations suggests further changes in the content of communications. Currently, we are experiencing the occasional use of temporarily united independent stations, called "ad hoc networks," for such premiere programming as biographical programs based on the lives of Golda Meir and Anwar Sadat, or even a dramatic presentation like *King Lear*. The success of these programs, plus the efficiencies of satellite distribution, suggests not only more elaborate and exciting additional programming, but also a strengthening of independent broadcast outlets.

At this juncture, however, the Premise of Mutual Success would alert us not only to the simultaneous success of pay television, which is also developing its own film and series product, but also to the increasing vitality of programming on individual municipal cable systems. The local independent stations and local broadcast network affiliates, like the national networks, will be forced to develop their individual strengths. Thus, a local broadcasting station, which can broadcast into the suburbs, might take advantage of its wide reach to present news programs emphasizing state and county issues; while a local cable channel with limited reach would do better to emphasize reporting of individual town events.

What the Premise of Mutual Success indicates for the broadscale media is a quantum leap in total product, with production quality spread out over all time periods and seasons and channels. It implies more aggressive, specific, and complete journalism, more widespread efforts at exciting and involving entertainment, a focus on more tastes and social strata, combined with epic efforts to attract everyone. It implies an increase in television viewing, up from a current 60 percent to 70 percent of households in prime time and from 30 percent to 50 percent in daytime hours. The broadcast networks will share the majority of this time, followed by independent stations and pay-television services, with the remaining 10 percent to 15 percent of the time divided among the ten cable networks provided on a typical local cable system.

The Premise of Mutual Success and Personal-Use Media

When we imagine the future of the personal-use media, most of us probably envision a personal computer occupying at least one corner of one room in our homes. Regarding the personal computer, we have already discovered that the filing, word-processing, and calculating functions, as attractive as they may be, will not have a significant impact on our lives. Yes, they will be there, offering convenience and adding a second literacy to our lives. And because of them we will grow more familiar with and adept at using home video screens. But the pattern established to date suggests that we will use these appliances mostly for entertainment and for only about half an hour a week on average.

The entertainment, informational, and facilitating functions of a home computer could be readily enhanced by its integration with videodiscs. At this point, it is too undisciplined to leap to the conclusion that such an integration will occur on a wide scale in the home. What makes the thought possible, however, is that the marriage of videodiscs and computers is being successfully arranged in malls to assist shoppers and in training facilities to assist educators. Based on these services, it is exciting to contemplate the possible programming that will be available on videodiscs for the home in 1990. A user may, for example, be able to select personal endings for a murder mystery or review statistical information from the Audubon bird guide and then watch individual birds flying, nesting, eating, and so forth.

Beyond computer interface, however, there will be an enormous programming library available on videocassettes and -discs. There is likely to be an aggregate of 35 million players in use by the turn of the decade, and each user will purchase upward of fifteen programs a year and rent twice that many. What these kinds of numbers represent is the foundation for the kind of variety and flexibility in programming alternatives that we now see in audio records from classical to pop, from poetry readings to foreign-language instruction. These will be added to the films and self-improvement exercise programs most popular today. The critical mass will be present for profitably serving the narrowest of special interests.

The success of MTV, the rock video music channel on cable, has already spawned competitive programming on other cable networks, on the broadcast networks, and on the independents. Hidden underneath this competition is the beginning of a scenario that will energize the sale of videocassettes and videodiscs while lowering their retail prices. It seems inevitable that this broad-based programming will perform the same functions for videocassettes and -discs that radio now performs for audio cassettes and discs. Sometime around 1990 consumers will be able to find whatever they seek in personal-use programming and software, at a price they can afford.

The application of the Premise of Mutual Success to personal-use media will naturally have a reverberating effect on some broad-based media. Film companies can now enjoy margins and profits from disc or cassette sales that parallel what they receive in theatrical release and that are significantly greater than their sales to pay television networks. This is also true, of course, for pay-per-view distribution. A. C. Nielsen projects that 35 percent of U.S. households will have a cable system offering pay-per-view programming in 1990. Nielsen argues in its study that if the ticket price is $8 and 30 percent of these households subscribe, the event would generate over $85 million. With only 17 percent purchasing a $4 ticket, the event would generate $24 million. As we have seen previously, projections indicate that 55 percent of the cable systems in 1990 will have pay-per-view capability. I tend to favor more conservative numbers for both pricing and viewing, but the net result is a similar estimate for potential monetary return.

At some point in the not-too-distant future these possibilities will create pressure on the existing pay-television networks to produce more original material and to lower the prices they charge. It is probable that at least one of the pay-television networks will pursue a video publishing strategy of lower direct consumer cost, balanced by a small percentage of the commercialization one finds on the broadcast networks. Pay-television networks will become advertiser supported as a result of competitive pressure from the success of other new electronic media.

The Premise of Mutual Success and the Print Media

Since we are concerned with the totality of communications, we must consider print as well as electronics, and discover what the Premise of Mutual Success might suggest about the future of books, magazines, and newspapers. Books, like radio, have encountered all types of competition over long periods of time and have survived and even grown because they fulfill their promise. Over 50,000 book titles are now being published annually, compared to 11,000 in 1950. The number of bookstores has more than doubled in the last few years, as have per capita book sales. Library circulation has consistently grown at twice the rate of population growth for the last four decades. Americans report spending about 10 percent of their free time reading, suggesting that books are an irreplaceable pleasure. Magazines and newspapers, however, are still evolving, and they will change as the new electronic media grow and prosper.

The easiest projections for the future of magazines suggests the current trends toward greater specialization will continue. Publishers will also discover new areas in American life that they feel warrant specialized coverage. In the past three years, almost six hundred new magazines have been introduced, with about 60 percent directed toward the consumer and 40 percent toward business subscribers. Many of these have failed, but there are also successes. The leading categories are computers, recreational sports, and new communications technologies. Within a few months of the introduction of the IBM PCjr personal computer, for example, eight new magazines aimed at this potential customer base appeared on the market. This trend toward specialization is also supported by the extravagantly growing list of very specific newsletters directed at people as diverse as chocolate lovers (e.g., *Chocolate News,* available for $9.95 a year), expense-account entertainers (e.g., *Restaurant Reporter,* available for $36 a year), and dreamers (e.g., *Dream Network Bulletin,* available at $18 a year).

This trend toward ever more narrow specificity on an expanding number of subjects will more than likely be balanced by broader-based magazines intent on interpreting and creating meaningful patterns out of the plethora of specifics. These magazines will pull together the pieces of our fragmented living, much as the newsweeklies, homemaking, and general business magazines do today. They will attempt to make basic sense out of burgeoning nonsense on social and domestic levels, but unlike some of the newsweeklies of today that now try to make sense out of everything for everybody, they will focus on somewhat smaller life-style issues. The proliferation of stories and news events will demand an essaylike format, for only a thoughtful point of view can throw a net over the multitudinousness so as to make it intellectually manageable and meaningful. Increasingly one point of view will necessarily exclude others, no matter how cosmic it intends to be. Thus the newsweeklies of the future may be an amalgam of *Time* and *Esquire,* a new kind of magazine yet to be imagined.

Many of these magazines will be supplemented by video

magazines available on both cassettes and discs. This is happening in trade areas today, such as fashion and advertising, where visualization is so important and a premium is put on personality. As television nudged *Life, Look, Collier's,* and *The Saturday Evening Post* off the newsstands or into different forms because television could perform many of their functions better, so too a growing base of video hardware will spur the creation of video magazines that will in turn change the character of printed materials, forcing them to place a greater emphasis on detail and analysis.

The Premise of Mutual Success and the Dialogic Media

The new electronic media perceived to be the closest to and most competitive with print are the dialogic media. Up to this point in our application of the Premise of Mutual Success, we have been able to build our picture of future communications on a foundation of the circumstances in 1990. The dialogic media, however, will only have modest penetration at that time and will not be in a position to have a competitive influence on the metamorphosis of other media until the latter half of the decade. At that juncture the success of the dialogic media will have a direct impact on magazines that are primarily information vehicles such as directories and buyer's guides. For example, *TV Guide* and *Consumer Reports* could transfer their listings and evaluations directly to videotext. The sheer quantity of information will be so great as to almost demand the efficiencies of available electronics. At the same time, however, this implies both greater value and much wider diffusion for such data and information than now exists, and the impact on society and how we make purchase decisions may be apparent long before the dialogic media have significantly influenced and altered the character of existing print vehicles.

The combination of local origination programs on cable television, covering community news and advertising, with videotext's ability to efficiently provide classified advertising and social announcements, goes directly to the heart of what newspapers provide today. To many minds, this combination of the new electronics dictates the demise of newspapers, but such predictions overlook both the power of print and the fact that current experience indicates that only about a quarter of Americans are interested in reading more than three consecutive paragraphs off a video screen. Under the Premise of Mutual Success, newspapers will survive, but they will change. The probability is that newspapers will become more like magazines with an ever greater emphasis on service editorial to assist us in conducting our daily lives and on making sense out of today's significant events through analytical reporting. This trend toward features and analysis is already apparent and more than likely will become the universal pattern of newspaper publishing in the future.

The success of direct-response advertising on television and cable channels today and the possible success for videodisc catalogues in the future is already apparent. In 1982 the telephone surpassed the direct mail of the postal service as a medium of direct-response advertising and transactions, generating $12.9 billion in sales. "Per inquiry" advertising, through which the communications medium is not paid for time or space by the advertiser, but in relation to the number of sales or leads generated by the advertising and its "800" number, continues to sustain the economic viability of several cable networks as it did during the start-up of Ted Turner's Superstation. Sears, Roebuck successfully tested a videodisc version of its retail catalogue in both stores and homes, and other catalogue companies expect to follow, as the videodisc player population grows to the point of viability. Not only is direct-response advertising growing by leaps and bounds, but all kinds of video media, joined with telephone and mail, check and credit card, are fueling a cultural change in shopping and purchase behavior. The phenomenon is growing independently of the dialogic media, and consequently our imaginings of the future should be premised on the success of these versions of responsive television, as well as on videotext.

The future of the dialogic media will not be in only one mode or as one massive webbing, like the telephone, which provides the same opportunities to everyone simultaneously. On the contrary, there will be great variety in national and local videotex and teletex services and the "closed user groups" operating outside of these networks, paralleling the very specialized small subscriber-based magazines and newsletters. People will have different combinations of horizontal and vertical dialogic media, not only as they do with print materials today, but also as they do with computer programs and services. At this point it is impossible to predict whether the dialogic media will be more like an encyclopedia or a library, a newspaper or a newsletter. We do not know whether they will be places where one looks more for brevity or depth, for information or transactions, for electronic mail or the downloading of video games and computer programs. Quite simply, it is too early to tell. What it is not too early to tell is that following the Premise of Mutual Success, the dialogic media will do all of these things, that they will support multiple businesses, and that consumers will tailor dialogic services to their personal needs and tastes.

The Emerging Video Landscape

All of the scenarios and hypotheses of the emerging video environment that we have generated from the Premise of Mutual Success have a mixed empirical value. They are limited in that they do not predict the actual content of our communications. Nor do they dictate what appliances any specific individual will have in his home. Neither do they project particular winners and losers among companies, products, and services. Their intended value is to provide a realistic, imaginative entry to the topography of communications. They outline the probable shape and general content of the electronic landscape. Their purpose is to provide us with a personal, existential feel for the future, rather than an abstract, dehumanized set of projections.

Importantly, these applications of the Premise of Mutual Success introduce us to the mutual dynamic of communications media and to their power over us as they compete for our time and dollars. These scenarios and possible futures allow us to imagine ourselves choosing, acting, and living in that important juncture where communications and culture meet. Each of us, however, must approximate our own future choices in the context of market conditions, penetrations, availabilities, and the research experiences explored in the previous chapters.

What any summary of individual media projections does not tell us is where and to what extent overlaps will occur. A simple application of mathematics suggests the obvious fact that 85 percent of cable subscribers will also be pay-television subscribers. Furthermore, 13 percent of these people will have home computers, and 7 percent will be utilizing videotex in 1990. When cable effectively reaches everyone in 2000, the vast majority of subscribers will also have three or four of the other new electronic media in their homes. Almost half the American population in 1990, and almost all in 2000, will have a minimum of two new media in their homes. One of the new media will be an entertainment system capable of satisfying any whim, and the other will be a random-access information system covering many subject areas and data bases.

In spite of the authority of tables and statistics, mathematics was never intended to be the best measure of human behavior. Nevertheless, what both the numbers structured on historical and business foundations and our working scenarios based on the Premise of Mutual Success do suggest is that *not all of us will have everything, but that nearly all of us will have a combination of new media in our homes that together will be sufficient to change our lives, behavior, and expectations.* The time-frame for such ubiquity and pervasiveness is fifteen years, a period not unlike the years following 1950 when television was suddenly omnipresent. We are only a few years away from such a generational moment, a time when high school students will never have known a day without these media, and when their young, vital, influential parents in the prime of life will already have consciously, willingly, and energetically reshaped their lives around them.

Equipped with all of the new resources we have described, how will Americans behave when they want to know something, learn something, enjoy something, buy something, or work at something? If they answer these questions in a manner in any way like what we have seen to date, they will be watching broad-scale television over sixty hours each week; they will add at least six hours of videocassette or videodisc viewing, and close to ten more hours with videotext and their home computer. All of this will be accompanied by no more than a 10 percent decrease in time spent with all other media such as magazines, radios, and newspapers. They will also purchase 30 percent to 40 percent of their goods and services through some direct-response medium. Of course these terribly generalized characteristics will change in proportion and composition from home to home, but under any probable set of circumstances *each of us will be spending upward of twenty-four more hours each week with communications in the emerging video environment of the 1990s.*

Where will all the time and product and resources come from? There are few satisfying answers. Communications, however, have always evolved incrementally, as in a spiral, and as we peer forward along that spiral, we can realistically imagine the simultaneous presence and use of the new electronics. We are now looking at a moment in time as significant as the emergence of the alphabet, the printing press, the telephone, and television. These new communications will inundate us with information and entertainment. Communications will be mimetically involving and quick as thought. They will be individualized and armed with the power of remote transactions. They are inevitable and will surround us, becoming the new tools of our culture. They will change our social character as assuredly as did television and the printing press.

PART 3

THE CONFETTI GENERATION

Having understood the two parts of the equation for predicting social character—people and technology—we can explore how their conjoining will create the Confetti Generation. We will examine the effects of the quantum leaps about to come in information, services, and entertainment choices, and the speed, isolation, and remoteness with which we will make those choices, and the necessity of having a personal survival kit to preserve our sense of value and personal significance.

The most general outlines of the coming communications revolution are clear. The basic socioeconomic trends encouraged by the new technologies have already been established. But the tools of personal exuberance, individual utility, responsibility, and self-creation seem both out of mind and out of reach. We must identify and fill this missing imaginative space, the discovery of the self in the future. By our own activity we can mold the new electronics to ourselves, or we will be molded by them in the most predictable, though not necessarily the most desirable, ways.

A "people" and the "masses" are two very different concepts. A people lives, breathes, and moves with a vitality of its own in the fullness of the individual men and women composing it. The masses, on the other hand, are inert and wait to be moved by impulsion from the outside. A mass works from the outside in, as many of us today wait for new technologies to alter our lives. A people, however, works from the inside out, as America itself has developed since its discovery. Waiting for and accepting the impact of new technologies would be ethically indifferent were it not for the fact that communications technologies change our private lives and personalities, and consequently what we do to one another.

8

The Confetti Generation

What will become of us, living our separate and collective lives in the topography of the new video environment? What developments in our social character might we reasonably expect over the next ten to fifteen years when the new electronic media become pervasive in our lives?

Our most personal concerns about the future spring from our encounters with one another, from the options and opportunities of communal living, the worldviews that come to dominate our thinking, and our general approach to the world and its people that evolves out of our daily encounters. We are, after all, less interested in our economy and its technologies than in how we will be living and on what we will ground our values and the goals we seek in our individual lives. Social character is the focus of our concerns about the future because it is the sum total of our behavior, the inner center of our culture, and the context of our personalities. Social character is the result of society communicating.

In earlier chapters, we examined the almost laboratory opportunity offered by the decade when broadcast television exploded in the midst of our other-directed world in order to discover the relationship between communications technology and social character. Today we are on the verge of a similarly critical juncture. The new electronic media, at some point in the not too distant future, will most assuredly change, modify, and define our social character, at least as much as the technologies of printing and television already have.

Our projections of the possible futures are often inhibited by our not having a practical and realistic understanding of the new technologies and what human life will be like in a society that uses these technologies to communicate. Often we are either dreamily indifferent or optimistically confident that the effects of technology are always beneficent. The last few chapters have tried to inject a sense of reality into our imaginings of future communications technology, encouraging an insistent practicality about what the new media and their products will be. In spite of this, our imaginations are still inhibited by the generalized images of our social future, which, for one reason or another we have grown to accept.

The Standard Scenarios

In general, there are three scenarios of how we will be living in the future. They all propose to tell us about our personal lives and about the options, possibilities, limits, and premises of our culture.

The first is simply an assumption that tomorrow will be an extension of today, but with more of it—more time, more money, more ways to satisfy our sense of self. This seems to be the working vision on which most of us operate—a vision rooted in the belief that the American way of life defines successful living, that the freedom to be you and me is built on financial achievement, and that those less privileged want precisely what the middle class now has. Incorporated into this belief is the assumption that the urge for more appliances, services, and leisure will fuel an economy with the scope and resources to raise the poor out of the cycle of poverty and support the services needed by the elderly. At the foundation of this projection of the future is the implied right and ability of the middle and upper classes to ever more convenience and pleasure.

These generalized expectations of the future assume that America has worked so far and it will continue to work. But because this hypothesis is essentially economic in nature, it is constantly buffeted by the cyclical upturns and downturns in the economy. When we are feeling pessimistic about the economy, we become anxious about America reaching the upward limits of its capacity. In prosperous times, the younger generations aggressively insist upon achieving the economic conditions and rewards they grew up to expect as the normal environment. They are impatient if their starting salaries are significantly less than the median family income they grew up with. Consequently, we might term this scenario the basic middle-class assumption, and it implies little change in our current social character. America will work, and we will continue to be who we are today.

Expanding this scenario and filtering it through a Gibbonesque view of history, we arrive at a second scenario similar to the one outlined in *Brave New World*. Here we see society rushing headlong toward greater self-indulgence and ever more concentrated self-gratification. We enter an age when our social character, driven by a mass-market hedonism, will be truly narcissistic.

A third scenario argues that this self-indulgent thrust of history will be interrupted by a major crisis, such as war or the depletion of our energy supplies. The resulting economic dislocation will demand sacrifice and social cohesion to the point that we will happily sacrifice our freedoms to a central power embodying socialism and government control. A less apocalyptic alternative, however, sees narcissism dying of its own weight as it confronts the limits of economic growth, with the result that we will all get back to basics, perhaps so far back that our social character will again become inner-directed.

Implicit in these and similar scenarios is either optimism or pessimism about the economy. As we review the American economy of the last forty years, we often overlook its remarkable strength. Even in the more recent ten-year period of growth, recession, and stagflation, America has shown tremendous resilience. In the last decade, for example, about 20 million net new jobs were added to the American economy, absorbing the increased numbers of working women, maturing baby boomers, immigrants, and alien workers. During the same period of time, the industrialized countries of Europe added no new net jobs. Unemployment in America has meant to this day that we are *not creating* 250,000 to 500,000 new jobs per month. This is not said in lack of sympathy—anyone's unemployment is 100 percent unemployment—but rather to emphasize the residual strength of the American economy, which takes body blows but recovers over time and, even when suffering, is dramatically better off than nearly every other economy in the world.

Though we can contemplate destructive and overwhelming crises such as war and national economic depressions, we cannot predict the time or place of their occurrence. Should such crises occur, they will create the need for a new formula for dealing with society's problems. But, until then, it is reasonable to make the assumption that the American economy and American institutions are strong enough and resilient enough to provide a satisfying standard of living for most citizen consumers over the next ten to twenty years. By eliminating both extraordinary optimism and extraordinary pessimism about the economy, we also eliminate the economically based projections of our social character. We must look elsewhere for a variable that might cause or influence changes in personal living. One possibility is technology.

When we turn to technology as the variable, however, the

first image that occurs to most of us is George Orwell's *1984,* which represents for many of us the most feared set of conditions for human living. Although *1984* is a book about the abuses of authoritarian socialism, and therefore a book about politics rather than technology, most readers (and nonreaders) are primarily disturbed by the relentless electronic surveillance. Chilling and unforgettable, the novel has consistently fueled the suspicion that any new technology will reinforce the power of bureaucrats and politicians, police and military, and either increase the powerlessness of the individual or hasten the onset of various forms of dictatorship.

On our mental bookshelves next to Orwell stands Franz Kafka, whose characters move not in a world of dictated truth but in a world where fixed truths are unavailable or impossible. Theirs is a world not of determined behavior but of random behavior and catastrophe. When these images of Kafka's random external world are associated with new technology, they are often linked with computer technology and then labeled Orwellian. (Ironically, Orwell never mentioned or imagined computers.) It is Kafkaesque images that come closest to paral leling our fears of impersonal, random, and meaningless quantification, and Orwellian images that translate these fears into a root fear of imminent totalitarianism made possible by technology.

It is impossible to walk around or away from Orwell and Kafka when contemplating contemporary life and future technology; and, depending on one's personal experience and politics, these writers may offer compelling pictures of the future. However, our fear of these projected futures is probably our best defense against them. And if we tried to map out the sequential steps that would take America to one of these futures, it would probably advance us beyond the ten-to-twenty-year time frame—an era in which it is extremely difficult to envision ourselves living. Thus without overlooking the prophecies of Orwell and Kafka, it probably represents a failure of our own imaginations to reach for Orwell and Kafka too often.

Ultimately, neither our standard scenarios for the future nor the compelling images of fiction adequately fill the lost imaginative space we now confront. We lack an existential feel for the near future, and an understanding of the vibrancy of social character. We cannot feel ourselves living and breathing in these scenarios. We need a fourth scenario, one less prejudiced by the preconceived values of fiction, a scenario informed by apprehension but less fearful of mankind, firmly rooted in past experience and grounded in what we know of the future of communications. We need a scenario to prepare us for the social character we may soon be living. For we are about to meet the Confetti Generation.

The Arrival of the Confetti Generation

Currently we are living in the Autonomy Generation. In following Carl Rogers's protocol, we feel that we are the center of all relevant values. We are responsible only to ourselves, and we alone can decide which activities and ways of behaving have meaning for us and which do not. We live subjectively according to our own feelings with little need for outside reference. In following Abraham Maslow's protocol, we interpret life in terms of what's in it for us, seek authenticity by transcending society and external value systems, and insist on being ruled only by the laws of our character. We claim to be spontaneous, uninhibited, and unpremeditated, and we live for ourselves, not for our predecessors or for posterity. We live in the present, responding to momentary perceptions, relationships, and encounters. To us, what is most important is how outside events are perceived and understood by the individual.

Relying on an internalized set of scales, the Autonomy Generation weighs personal and social values: in one dish is who we assert we are today, this week, this season, this year; in the other dish are weighed all other experiences in an almost undifferentiated flow. Only those experiences that temporarily balance are temporarily accepted. The Autonomy Generation is suffused with the notion that there is no objective reality, only individual experiences of it.

On the surface it would appear that our current Autonomy Generation must suffer from what Durkeim called *anomie*, the peculiar pain derived from individuals' inability to identify and experience their community. Such group anguish hasn't developed in large measure because of communications. For all the ability of individuals to pick and choose among communication products and to interpret these products in individual ways, the fact remains that broadcast television's ability to reach everyone and satisfy everyone has enabled society to maintain something of a common experience and common agenda. It has proven to be a strong connective tissue in the body politic. The Autonomy Generation has not in fact broken the fabric of "society communicating," largely because of the communications technology available, the interconnecting nature of much of our work, and the manufacture and marketing of mass consumer goods. The question remains, however: What will happen when this generation encounters the general availability and pervasiveness of the new electronic media over the next ten to fifteen years?

The combined new electronic media share several general characteristics that will affect the way our society communicates, the base for our culture. The first general characteristic the new media share is *quantity*, in both availability and use. Not only does a cable system provide more than six times the number of broad-scale communications channels than were previously known to the television audience, but the new personal-use and dialogic media create a whole new constellation of available information and entertainment. The average American will be spending twenty-four more hours each week in communications consumption.

The second characteristic of the new media is *speed*, both in delivery and satisfaction. Already we are accustomed to news being broadcast through radio and television almost before it happens. The new electronic media will continue this trend, feeding upon constant and ever-changing information like a stock ticker spewing forth millions of financial transactions as they occur. Editorial selection and the winnowing-out process it implies will collapse. There will be no gap between impulse and satisfaction on the reception end, whether the consumer requests hard information or mild entertainment. From our individual standpoint, we will be living in a world of "quick in/quick out" and "quick out/quick in."

The third characteristic is the *weightlessness of images* that are the capsules and carriers of information and entertainment. On the one hand, television images are intensely mimetic, and thus rapidly internalizable. On the other hand, these intensely mimetic, rapidly internalizable images occur in their own world and without context. The images of events *become* the events, and their rapid dissemination robs them of a past, insists on the present, and excludes the webbing of perspective. Furthermore, all information, whether sporting event or daytime serial, video game or spread sheet, commodities report or philosophy, occurs in the same electronic framework, on the same video screen, without differentiation or distinction. The result is a visual homogenization of the value and significance of images.

The fourth characteristic of the new media is *remoteness*. We are all familiar with broadcast television's ability, when armed with satellite technology and supported by the power of other media, to bring us into immediate contact with the remotest reaches of earth and space. Because of these technologies and services we now live in a horizonless world. Global events affect us, and we are led to believe that we affect them. But consider the remoteness implied by the new electronics. We will hold in our hands a remote switch empowering us to engage in remote shopping or remote banking. However, this very possibility implies its reverse. Remote merchants and bankers can find and deal with us, and they need not see us anymore than we need see them. Thus the new remoteness implies both isolation and the ability to isolate.

The fifth and final characteristic is the fact of *choice*. The enormous growth of communications, the explosion of available information, entertainment, and transactions will, in fact, necessitate and compel more frequent and more personal choices. Limited choice induces homogeneity in everything from politics to shopping, from evening meals to the use of leisure time. Multiple choice induces heterogeneity, and unlimited choice induces individuality. Thus the final and perhaps overriding significance of the new electronic media is the fact of choice they will continually impose on each of us.

When all five of these characteristics—quantity, speed, weightlessness and remoteness of images, and the fact of choice—come together in the new electronic media, the ability to fracture the community of experience and break the historical patterns of society communicating will produce the Confetti Generation.

The new media will cause a new kind of individualized cultural segmentation. The tools will be the quantum leaps in available consumer information, in sight, sound, and motion entertainment, and in speed and facility, as well as the isolation and remoteness of transactions made possible by the new electronics. The new electronic media will be alluring, eagerly adopted by almost every element of society, by the impressionable young, by macho achievers and their emulators, by intellectually integrated adults. Each consumer will have his own motivations, but together everyone's actions will effect a sudden change in social character, a change paralleling the transformation of social character brought about by the invention of printing.

The result will be the Confetti Generation, for the current Autonomy Generation does not possess the cultural tools to absorb such an explosion of information and entertainment, such an implosion of speed and remoteness. Having been nurtured in an Autonomy Generation, the Confetti citizen consumer will be inundated by experience and ungrounded in any cultural discipline for arriving at any reality but the self. We will witness an aggravated version of today when all ideas are equal, when all religions, life-styles, and perceptions are equally valid, equally indifferent, and equally undifferentiated in every way until given value by the choice of a specific individual. This will be the Confetti Era, when all events, ideas, and values are the same size and weight—just pale pink and green, punched-out, die-cut wafers without distinction. The new electronics will create the opportunity for eclecticism to run rampant, an opportunity for such increases in speed and satisfaction, information, entertainment, and transactions that one will be almost forced to select randomly and remotely based on personal, fickle taste. All the while, the sheer weight of data acceleration will create the impression of similarity and equality of value. Social character is produced from choices, and when ideas and experiences float down like cheap confetti, the Autonomy Generation will choose in the only way they know. By their choices, they will become the Confetti Generation.

The Confetti Generation at Home

The question I have most often—even consistently—been asked at seminars throughout the world, in cities as disparate and different as Acapulco, Amsterdam, and Atlanta, is: "Will the new electronics, centered in our homes, bring our families closer together?" The premise of this question implies a consensus on a contemporary problem for the nuclear family and a certain hope. Since the new electronics are centered in the home and enable us to satisfy so many of our needs from the home, perhaps families will find themselves in the home more often, gathered together in the same room, and therefore acting as a more cohesive and interrelated family than they do today. Technology, however, does not create a family, nor do the increased physical conveniences of the home necessarily improve the life within it. What the question does emphasize is that our key concerns are personal and domestic, that we will see the impact of technology in our relationships, and that the pulse of the Confetti Generation will be felt in the home.

To imagine family life in the Confetti Generation we must begin with current trends. In the last decade the number of individual independent households has increased at twice the rate of the population as a result of later marriages, dropping fertility rates, the number of divorces, and the increase in longevity. Some of these trends seem to be modifying, but the overall impact is that on the average there are fewer than three people living in a typical American household. The percentage of working women, educated white-collar workers, and the more affluent is growing greater every year. These are the base factors influencing both the acquisition and consumption of the new electronics.

Americans still invest heavily in home appliances. That is why sales of videocassette recorders have so directly paralleled the early sales curve of color television sets, which rapidly made them the typical home receiver. It also helps explain the rapid growth in multi-set households. Now more than half of our homes have at least two television sets. The number of homes with more than one set attached to cable television is also growing. Not only are home computers equipped with their own display terminals, but in multi-set households, they are often attached to separate television sets (other than the VCR or cable sets.) Thus our vision of the future should include different television display devices attached to different information and entertainment appliances in different rooms, and utilized by different people seeking different services at the same moment.

It is true that today many families happily report that they now sit down together to watch a movie from pay television or from a rented videocassette as they used to watch programs in the early days of television. But it is unlikely that they will continue this practice any more than they did with broadcast television. In the experimental tests, families tended to spend hours together with videotex, but since videotex is designed to serve very disparate needs, it is hard to interpret this family watching as anything other than the experience with a new toy, and it is unlikely to continue.

Thus if the new electronics are successful, they will be successful on their own terms by serving specific needs and gratifying very separate tastes. Overall, we have to imagine the segmentation rather than the sharing of experience, both in more separate households and in more separate uses within each household. The facts, trends, and characteristics of the new electronics do not support the dream of more togetherness in our homes.

The dynamics of consumption—the relation between the communications product and the consumer—also are likely to support feelings of isolation. The information products within our equation will grow increasingly insignificant and incomprehensible even as they inundate us with sheer volume. An excellent example of one woman's personal response to this inundation was provided by Meg Greenfield in her *Newsweek* column the last week of 1983, in which she argued that "this was the week when the foreign news finally became totally incomprehensible." She continued:

You think I'm being frivolous or hysterical or both. Very well, I will tell you exactly what the daily news take had been at the point when the big board lit up. The Israeli Army had helped to get 2,000 "Lebanese Christian militiamen" out of harm's way, one learned, by evacuating them in a convoy of 100 Israeli trucks to the coast. The battleship New Jersey had shelled "Syrian-backed Druse militia positions in the mountains." This was in retaliation for a Druse shelling of a Marine encampment that earlier "had come under heavy fire as the Lebanese Army's artillery batteries nearby clashed with Shi'ite Muslim militias near the U.S. and British peacekeeping zones." Meanwhile, Greek ships, under U.N. flag, were preparing to evacuate "PLO leader Yasir Arafat and his loyalist fighters." That story went on: "The fighting in Tripoli today was between fiercely anti-Syrian Muslim fundamentalist militiamen—who had fought alongside Arafat forces in the last month's intra-Palestinian battles—and their longstanding pro-Syrian Alawite militia foes."

You might as well give up: nothing you can say will convince me that you understand any better than either the American government or Pundits Inc. does who all these various warring elements are, let alone what they want.

Meg Greenfield is an editorial writer and may be suspected of having an editorial bias about Mid-East affairs, but I believe she is describing a problem most Americans would admit to in relationship to news reports regarding many parts of the world, or to the competing arguments about our economy. Most of us would acknowledge this kind of problem with the evening news on any given night and certainly to the news taken on consecutive nights. New technology in all of its forms will simply aggravate the confusion. Information will rain on us like confetti and become just as meaningless. The information we receive, isolated with our television sets, will be increasingly incomprehensible.

Incomprehensibility derived from information overload is one description of our future communications product. To this we must add the style and character in which these images of information will be presented. The style and character of presentation will not be much different from the trends we can observe today. At the glossiest end of the spectrum stand the new music videos popularized by MTV. Their fractured forms, thrusting gestures, and frozen frenzy are characteristic of much modern visual art, but in art the elements relate to each other and the images cohere. This is not true of our popular video products or printed graphics. Both magazines and television today serve up serious, emotion-laden information with all the color, frenzy, and fractured gestures of contemporary visual art without any necessary coherence.

These crazed images and the stories they tell apparently have no roots in reality. Most of the magazine articles and stories today, and even books on the best-seller lists, are far glossier than life. It used to be only in some Hollywood movies that you couldn't tell what kinds of jobs people had—or how they could afford the houses they lived in with the jobs they had—or how they could be concerned about the silly things they were concerned about. A scan of current products would suggest that we are more concerned about the shape of our bodies than about our children, and more concerned about vacations than losing our jobs. It's as if print had gone Hollywood.

We should be shocked by the way business is conducted on the slick evening serials such as "Dallas," not because of the immorality or amorality or amounts of money and sex, but by the way they distort manner, speech, detail, and time in the depiction of business deals and strategies. There is also a problem in much television programming of oversimplifying complex moral and social issues, such as incest or life-prolonging medical ethics, into narrow and superficial personal stories in which the ultimate risk or resolution is no more difficult than traffic court. Our increasing consumption of fantasy and science fiction is certainly escapist, but the escape is generally into a very hierarchical and simple society not the least bit reflective of or analogous to our own.

What the rhetoric, fantasy, unrootedness, and unrelatedness of much of our communications images have in common is not what is put into them, but what is left out. They are as light as confetti. Speed is what energizes these images: the speed of their arrival, the speed of adaptation to fleeting tastes and passions, the speed with which the images jump from one to the other, and the speed with which we jump between them. Thus the problem with this type of information overload is not simply quantity, but the unconnected, excited nature of the images that package and distribute this information, whether it be news of the world or stories of human interest. The ultimate result will be communications images that blow around in our homes like confetti in a storm, and just as randomly. There will be no apparent pattern.

Information from multiple sources, whether news or entertainment, is not required to have a pattern. In quite separate households in front of quite separate display terminals, the people choosing and receiving this accelerated flow of information will be our current Autonomy Generation. Encouraged by Rogers and Maslow to be our own exclusive reference point, and encouraged by McLuhan to find our own individual and unique patterns in the images of experience, we insist that reception is a process of eclectic free association. We have totally private agendas and are convinced that our desires create relevance and history. Metaphors and images are meaningful if we say they are, as images flash by us like words on so many flash cards, connected and interpreted as a kind of private Rorschach test. Our Autonomy Generation insists on its own lucidity, and since we are anxiously uncomfortable with ambiguity, we replace passing doubt with passing conviction.

Thus there will be a growing similarity between the new technology and the people receiving it. The speed and inundation of unrooted images will be met by an audience prone to speed reading video images, the way they have learned to speed read print. The trick is to scan with a clear idea of what you are personally looking for, to scan the headings, graphs and charts, pictures and captions, and, assuming that the text is well written, the first and last paragraphs. We are already living with these kinds of video reading habits, now encouraged even more by the computer, which is itself a speed-reading instrument. A person scans graphics and brief paragraphs of text without any understanding of the computer's inner logic of selection and presentation. Consumption of these images as such represents an acceptance of the surface of things as offered, as well as an intrinsic attitude of "pick-up and discard" in the process of creating one's own relevance and logic. In short, the confettilike inundation of images of information will be received by people who have proved themselves ready, willing, and able to generate their own confetti from even the most serious substantive images of art, history, and analysis. Such an equation does not represent the grounds of increasing togetherness in the home or anywhere else, but something quite the opposite—the Confetti Generation.

The Confetti Generation at Work

We are often told that in the future much of our work will be "homework." We will be "telecommuting" from our "electronic cottages." This is an appealing idea. As we mentally rearrange the furniture, we dream of being liberated from all the things we don't like about the office—commuting, bosses, gossip—and we are stirred by all the things we like about our homes—spending more time with the people we love, having flexible self-managed time, being in a comfortable environment. The idea appeals to parents who see it as a solution to the tension between work and child rearing, and it appeals to those who simply like the technology of it. It even appeals to certain company managers who think of all the overhead expenses they'll save in office space, lighting, and heating. Amid much of this speculation, however, little thought is given to when we will be able to work at home and for how many of us this will actually be possible.

People have always done business from their homes, from knitting to selling, from accounting to writing. However, there are only about five million home-based businesses today, and most of them have very little to do with communications technologies. More people are working at home because of their entrepreneurial spirit than because of computers. New technologies have added to the number of people working from their homes, but if they were isolated out of the total number of home workers, their number would be very small—maybe only ten to twenty thousand. This is hardly sufficient indication of the definitive trend people are expecting over the next ten to twenty years.

There are also clear lines of resistance developing. Company managers, reluctant to let workers out of sight, will seek to monitor home workers electronically or trust only a few very highly motivated employees to work out of their homes. Unions fear a return of exploitation, from the imposition of piecework wages to the return of child labor. And a lot of people discover that they lack the necessary discipline to work at home and that they miss the human contact of the office, while tensions, rather than togetherness, enervate their families. While it's true that more people will be working at home, it is doubtful that home working will become the fundamental pattern of our working lives by the year 2000. Nor will many of the home workers be dependent on the new electronics.

Most of us, however, will be working, and the nature of our work will continue to be critical in forming our social character. Few of us would disagree with the National Academy of Sciences' conclusion expressed in a 1979 report that "the modern era of electronics has ushered in a second industrial revolution [whose] impact on society could be even greater than that of the original industrial revolution." In the rhythmic phrases of General Electric's James Baker, there is general agreement that industry must "automate, emigrate, or evaporate." The last two alternatives are much the same, for they mean that industrial production won't exist in America. To prevent that, we will automate, and to a great extent that means robotics. The workers still on the line, controlling and monitoring these electronic and computerized production processes, will be fewer in number and will experience a revolutionary change in their involvement with their work.

Computers will design tools, consumer products, and even words. Computers will control production processes, whether at foundries or on printing presses, and they will control distribution from warehouses to electronic mail. Individually and in combination with each other, computers will eliminate the skills required for some jobs and increase the skills required for others. Microelectronics will change the way we do our jobs, and will alter the shape and vocabulary of our work. This change in style will affect our social character. Most critically, the number of people we must talk to and rub shoulders with will be decreased, and the touch and feel of hardware and paper will be replaced with video-display monitors and computer control panels. There will be no more heavy lifting, whether we are dealing with things or data, and there will be a lightening of the freight of human encounters, plus the judgment and inter-personal skills they require.

While most of us now complain that we are so overwhelmed with data that it is hard to see the forest for the trees, major managers are happy that at levels lower than their own they now have data with which to identify and monitor the processes of business. And few complain that they now have data and programs to use in circumstances where previously they were dependent on the judgment and experience of their staff. White-collar work today, as assuredly as the production line, has been broken down into discrete steps. This is one reason why there has been such an increase in white-collar workers.

The new problem, however, is that with the aid of microelec-tronics and computer programs embodying the latest in man-agement and marketing sciences, there is less need for elaborate staffing. All it takes is a recession to prove it. In the past, most middle-management jobs have been recession proof. But not only do middle managers now get laid off during economic downturns, just as factory workers do—their jobs often disap-pear when good times return, as is also true for factory work-ers. In this way, white-collar and blue-collar workers alike are suffering the effects of automation.

Fear of computers, coupled with concerns about job security, is leading many middle mangers to resist automation as strongly as factory workers are. But their efforts will be unsuc-cessful. The workplace is changing. Those retaining their jobs find that raises are neither as large nor as automatic as they once were, that they are working much more intensely than before, and that promotions are harder to come by than they used to be. In short, not only the opportunities, but the dynamics of white-collar work have changed.

In *The Lonely Crowd,* Riesman argued that the central problem of business in the 1950s would be people management in large white-collar organizations; he said that this would lead to the pressures of sociability and require "the work of men whose tool is symbolism and whose aim is some observable response from people." These new workers would use communications as a tool of sociability and turn to mass media to discover models of behavior and approved images of style, taste, and language. To get along, one had to go along, and Riesman accurately predicted that successful living in this environment would produce an other-directed social character. The inner-directed social character that preceded other-direction was also keyed to communications and work. On the one hand, it was a response to the proliferation of experience brought about by the printing press and mobility, and on the other hand, it was a means of ensuring responsible industrial workers keyed to the accumulation of wealth.

Similar changes in work and communications brought about by the new electronics will also alter our social character in the future. Other-direction resulted from a shift from concerns with *things* to concerns with *people,* and the question now becomes: What will be the result of a shift in work and communications from concerns with things and people to a total involvement with electronics—with light panels and video displays of letters, numbers, and graphs, with information articulated as data, with things and processes described in alphanumeric symbols, and with isolated, weightless experience on a cathode-ray tube? As in the past, we will become what we do and what we know.

The image of us all working from our electronic cottages contains a basic truth about our future work: We need not be there anymore. What the image implies is that tasks will become so much more narrowly specialized that they can be done without supervision. More important, it implies that these tasks of interfacing with information on a display screen can be done remotely because they are mechanical and anonymous. We'll be in the office, not our cottage, but we might just as well be anywhere—or nowhere.

Work will become isolated, intangible, and intense. Isolation will result from the reduced number of workers in any area, from a reduced need for face-to-face communications, and from the "me and my machine" syndrome already prevalent among current computer jockeys. Intangibility will result from the very nature of displaying and digesting computer information and from the remoteness from things, processes, events, and people represented on the display screen. Intensity will result from the specialization and repetition of tasks designed to satisfy a demanded efficiency and from the speed and transformation of data input. Thus there will be a lessening of peer encounter and an atomization of our attitudes toward work. We will be more concerned with data and process than with products and people. The familiar "organization man" will become a "process man." Instead of being participatory, we will become peripheral. In previous generations, changes in work and communications have transformed us from inner-directed to other-directed to self-directed. In the next generation, changes in work and communications will produce yet another social character—the "undirected." And the undirected are the people of the Confetti Generation.

The intensity, isolation, and intangibility of our work will make work less personal and less of an experience base for living. The nature of this work will also have a major impact on our seriousness about work and on our paths to success. Both will be reflected in near-continuous training.

Today it is not uncommon for people to spend as much as three months a year in some kind of training. In the future, we will need even more training for both business development and personal success. As training becomes more prevalent as a business way of life, it will become the arena for impressing one's peers and management. At the same time, continuous training reduces one's identification with a current job and even future jobs, since the intention will be to get out of each particular job as fast as possible and roll it over into another job with yet further training. Simultaneously, the economic rewards of each step will be marginal. The overall results of this dynamic will be a perception of work as valueless, a separation of work from our lives, a search for self-satisfaction elsewhere, and a new insistence on self-promotion as a means of grasping identity from anonymity in any way we can. That will be the Confetti Generation.

The Confetti Marketplace

The marketplace for goods and services is an arena where society communicates and where our social character is articulated and reinforced. When John Kenneth Galbraith concluded in *The New Industrial State* that our current mass market is "profoundly dependent on broadcast television and could not exist in its present form without it," he not only identified the intrinsic relationship between mass communications and mass marketing, but also the intrinsic relationship between communications and marketing in all their forms. Any change in communications produces a corresponding change in marketing. Mass marketing reflects a large and relatively cohesive society supported by extensive and relatively homogeneous communications. Without such communications, the mass market and the cohesive society it reflects will necessarily fragment.

As consumers we are reasonably conscious that what we buy is an articulation and a reflection of ourselves. For most of us this is obvious in our purchases of furniture, clothes, home decoration, automobiles, and entertainments. It is also true of what we buy to eat and how we prepare it, whether frozen foods or nouvelle cuisine. We luxuriate in the choices we have in an affluent America, from supermarket aisles to travel agencies. We have dramatically more choices than we had thirty years ago, and myriad more ways to express ourselves and reinforce our style of life. Yet compared with future possibilities, today's choices are limited. Almost every element of the product development, production, and marketing process will become ever more microspecific in the years ahead.

The marketing process begins with the identification of consumer wants, needs, or expectations, and there is little doubt that advances in the art and science of consumer research have introduced, in the last decade, more products more finely attuned to consumer wants and expectations. Consumer-research methods will continue to improve the ability to identify ever new and more narrow markets for ever new and more specific products.

Two limitations on successfully refining consumer research have been the cost-efficiency parameters of the production, packaging, and distribution process and the effectiveness of available communications vehicles. The new electronics have lifted these two limitations. The productivity efficiency of the new technology, plus the flexibility it affords on the production line, will enable marketers to efficiently produce and distribute an extensively greater product line than they can today. Unquestionably the new media will enable marketers to reach smaller and more dispersed audiences of consumers in a continuous, cost-effective way. The result of these combined developments in consumer research, production, distribution, and persuasive communications will be an extensive demassification of marketing. In turn, this demassification will contribute to a fragmentation of society communicating.

Two basic principles for imagining our future social character emerge from these combined developments in marketing and communications. First, the greater the number of distinctive consumer clusters that can be identified and satisfied, the greater is the potential for demassifying the marketplace. Second, the more these separate, distinct, and dispersed consumers purchase and use products designed for their own specific self-projections, the more their separate characters will be reinforced. At its roots, therefore, marketing has the power to form and reinforce either greater social solidarity or greater social fragmentation.

Tempering an exclusively marketing point of view is the fact that the number of consumer clusters that would normally develop and the number of clusters that marketers will be actually interested in will be rather small and evolutionary in character.

Mass marketing will continue to be profitable and act as a brake on fragmentation. Consequently, strictly market-derived changes in society could be expected to be reasonably modest. But if, on the other hand, there are other forces energizing a more rapid fragmentation, the marketing of products and services will change more rapidly. In the decades ahead, marketing and technologically driven social change will converge and energize one another in a rapid revolutionary spiral. It is precisely the changes in citizen consumers brought about by the new electronic media that will drive marketing into radically new fragmentation and segmentation. Marketing and new communications technology will thus combine to create the Confetti Generation—a group whose very eclecticism, fickleness, and random choices will demand a proliferation of products to reinforce their self-declarations. Demassification will verge on atomization.

When these marketing and communications factors are combined over the next decade, it is reasonable to project that products will almost inevitably have shorter life cycles. In many areas, brand loyalty will become a fond memory as the Confetti Generation becomes more conscious of picking and choosing among multiple products and as manufacturers trip over one another to satisfy the latest fad. In general, there will be an emphasis on more personalized products that are not simply variants or line extensions of well-known brands. Every home will be eccentric, and eccentricity in the purchase and use of products will become the norm. We will be faced with living the rapid formula: disposable income plus disposable life-styles equals disposable products and product loyalties. The more fragmented and specific this dynamic becomes, the more rapidly will we become the Confetti Generation.

The Confetti Mind

During the Confetti Generation the marketplace of ideas will become similarly fragmented, segmented, and remote. In religious thought, for example, theologian Harvey Cox points out in *Religion in the Secular City* that Christian fundamentalists "have gained great influence in recent years through their imaginative use of electronic media." It is ironic, he suggests, that "the fundamentalist electronic preachers have turned the religious congregation into individualistic consumers of a mass produced religious commodity." These new media evangelicals are particularly sensitive to the charge that their video services draw people away from local, indigenous congregations. For our purposes it is important to observe how the communication works—how by reaching one person at a time through discrete media, a community of active interest can be created among otherwise remote and isolated individuals.

Much the same thing is occurring in politics. Fund raising through direct-response advertising has proved very successful in crossing, even obliterating, normal geopolitical boundaries for liberals and conservatives. Almost all politicians now use zip codes representing particular demographic and psychographic groups to identify potential clusters of support and plan their appearances only in these zip codes while ignoring and avoiding others. Of course, with the aid of sophisticated polling, their messages are equally and discretely tailored to these clusters. In 1980 a *New York Times* editorial described John Anderson as "the first presidential candidate of what might be called the Television Party." Since then there have been any number of candidates who were introduced on television as a new product with no history and no foundation of support in any particular community or geographically cohesive constituency. We now have the ability to vote with our checkbooks or credit cards, and soon we may have the ability to vote directly from our homes through electronics.

The Confetti Generation—isolated, remote, segmented, and fragmented—will support an eclecticism and fickleness in the marketplace of ideas in the same way and for the same reasons as they will in the marketplace of goods and services. Both the purveyor and receiver of ideas will exist in a direct relationship that bypasses the moderating influences of a local community. In time, there will be purveyors of ideas with a radically more restricted view of society than can now be successful, and their consumers will have less reason to be thoughtful and prudent. Thus electronic communications, which have united us over the last four decades, can fragment us in rampant parochialism. More inclined toward anarchy than toward totalitarianism, the Confetti Generation will bring an age of rhetoric, an age of ballyhoo promotion for consumer goods, an age of demagoguery in the marketplace of ideas.

Destination and Destiny

To appreciate the living experience of the Confetti Generation, we should ask ourselves a few simple questions. When ideas and experiences float down upon us like confetti—and just as cheaply—how do we expect to choose? When the day arrives that we are rapidly and increasingly inundated with information, how will we sort it all out in our heads and in our lives? More important: When that day comes will anyone else we know, or anyone else within our community, share any common ground of experience and information with us? Will they sort out the data and opportunities in any way similar to our own?

In view of what we now know about ourselves and the future of communications, we might well conclude that without some new set of cultural tools, the evidence upon which we will structure our judgments and behavior will be unique to ourselves, and that there will be little common interpretation of the experiences we share. These will be social conditions we have not encountered before. It is the extreme opposite of the tradition-directed societies that preceded the first industrial revolution or the organized, other-directed society that preceded our own.

The electronification of our lives that began with television and the use of computers in the 1950s and '60s will become pervasive and microspecific sometime in the 1990s. That is our destination, but what is our destiny? To some it may appear that life will be different only in that it will be more convenient, but major changes in communications and work have always brought about major changes in social character. It is likely that this will happen again. In the Confetti Generation, we will be living undirected lives and life-styles. We will think and act not only for ourselves, but by ourselves, isolated from the common experiences and judgments of others. Such singularity of perception and action has never been possible before. We are about to think and choose and live differently than ever before in history.

9

Changing the Equation

The Confetti Generation is inevitable. It is where we and technology are headed. On one side of the equation is communications technologies with their own inherent characteristics and inescapable futures, moving in a fixed orbit, altering our lives at a steady pace. On the other side of the equation is we, the users of technology and the creators of communications. We are the variable in the equation, what we now think and do, and what we *will* think and do. Because of the inexorable advance of technology, it is unlikely that we will have either the time or the will to change ourselves before technology radically changes the conditions of our lives and choices. By then the Confetti Generation will have arrived.

Any viable prediction of the future must be rooted in a candid and forgiving recognition of who we are. Most critics of our society, culture, and social character in the past decade have condemned the narcissism and self-concentration of our Autonomy Generation. Most of their criticism, however, reflects a nostalgia for a preelectronic, inner-directed past, as if they would urge us into living a revised kind of Protestant ethic.

There is a great deal of historical and biographical evidence that in times of confusion and crisis, people and societies have a tendency to reach back over an entire generation of values for a renewal of confidence. In seeking guidance, we skip over our parents, as it were, to the perceived image of our grandparents. Something like this seems to be going on in much of the current social criticism that calls for a renewal of that old-time religion, asking us to revert to a set of values reminiscent of an age before television. Understandably, we refuse to go back.

The inner-directed conformism of a pretelevision age is simply beyond revival in light of the educated affluence we have developed on such a large scale. We have learned too much, enjoyed too much, and expect too much to return to the fundamentals of inner-direction. It is equally impossible to imagine a new strain of inner-direction controlling the imminent wave of new communications technology in quite the same way traditional inner-direction controlled the impact of printing and literacy in the past. We will not develop a sense of collective guilt—or even a consensus on which to base the guilt—in time to control the direction we are headed.

In these times of change, we are remarkably resistant to changing. When we think back over the years since the arrival of television, we are confronted on a national level by assassinations, Vietnam, riots, and Watergate. On an individual level we recall changes in jobs and locations. On the level of personal relationships, many of our histories are framed by marriages, divorces, and remarriages. In the context of these memories, it is easy to believe that all that matters is what we face today and what lies ahead in the immediate tomorrow. We have more reasons to forget than we do to remember, and too many personal and social disappointments, to cast our hopes strongly on the future. And it is this very belief in the moment that makes it so difficult for us to change, as much as we seem to change on a daily basis. We cannot change without a memory and a goal, and both are defeated by our lack of confidence in anything but the moment.

In theory, we can change. But practically, we will not change in the significant ways required to delay or avoid the development of the Confetti Generation. In some respects we are blind to the consequences of the full impact of the Confetti Generation. After a period of frenetic enthusiasm for the new technologies, the Confetti Generation is liable to develop a deep psychological depression bred of inner confusion and weightlessness. Burdened by speed, intensity, and remoteness in our communications, we may well yearn for the tyranny of minimal order—any order. Out of desperation, we may produce an insistent interior fascism that insists on its opinion beyond persuasion. Some may perceive this as a healthy return to inner-direction, but, in fact, it will be an interior response to chaos, paralleling the conditions and demands that produce social and political fascism.

Neither confusion and weightlessness nor a new interiorized fascism are happy prospects. To avoid these stark alternatives, we must ensure our sense of value and significance by creating a personal survival kit—one that does not restrain either technology or ourselves. Its utility lies in equipping us to monitor *how* we do what we do rather than telling us *what* to do.

The practical tools in a personal survival kit must assist us in absorbing, selecting, and rejecting the constant and overwhelming flow of images generated by the new electronics. At the personal level, they must directly address the increased speed and remoteness generated by these new communications technologies. The way we conceive of communications is fundamental in this process, for communications of all kinds are a means of knowing and experiencing. In a media-saturated society, we must ask: What is reliable knowledge? How can we acquire it? What can be done with it? A media-dominated environment switches the focus of our concerns from external to internal considerations, from information as generated by the new media to what we will believe and use as knowledge. The new arena for survival is the personal imagination that will deal with these questions.

We must explore the process of managing our imaginative life and apply it to the inundation, speed, and remoteness of images inevitable during the Confetti Generation. What we expect of communications on the social level must be consistent with what we expect and how we execute communications on the personal level. Especially during the Confetti Generation our social character will ultimately be dependent on our personal responses and responsibilities. It is our conception and use of communications technology that will ultimately make the difference, determining the media's products and services on the one side and our social character on the other. Our practical future, therefore, depends not so much on who we ought to be, but on how we do the things that make us who we are.

The Scientific Fallacy

During the Confetti Generation, and more than at any other time in history, our imaginations will be the battlefield of our struggles. Unfortunately, Americans have very little faith in the imagination as a faculty of knowing things, distinguishing things, and usefully combining experiences.

It is part of our cultural heritage as Americans to believe that the physical sciences represent the only discipline for understanding objective reality uninfluenced by the observer or reporter. We have generally concluded that, outside of the sciences, everything else in life is subjective reality, a matter of semantics and personal opinion. Consequently, when it comes to the life of the imagination, we allow ourselves to get caught up in the surrealism of the moment, either oblivious to the power of images or assuming that we are immune to their influence.

Our personal survival during the Confetti Generation—our ability to maintain a sense of values and personal significance—will depend in large measure on how we select, reject, and combine the images brought to us by technology. Our imaginations will be the arena for survival or despair. If we expect to have any hope in such a future, we must accept the fact that the images we create and absorb, the stories and symbols we give our attention to, affect us as human beings whether we intend them to or not. We must accept that these "pictures in our heads" affect the stand we take in the world, our perception of ourselves, our relationships, and our behavior. When the new electronics inundate us with images at cyclotronic speed, we will take them seriously only if we believe that our independent survival, our self-control, and our conscious self-creation are at stake.

We emerged from the long history of Western thought, so far from Plato through Descartes to the present moment, with six assumptions about nature and our personal lives: (1) that whatever is scientific is objective, and whatever is objective must be scientific; (2) that science is a process of quantifiable observations and mathematical expression; (3) that although science is a search for knowledge, it is more reliable as technology, as a tool for controlling of our environment; (4) that all else is subjective, a matter of opinion and convenience; (5) that the imagination is a faculty of entertainment, harmless fantasy, mythic memory, and personal projection; (6) that life is undiluted process without objective or scientifically discoverable ethical norms.

The great historical arguments about science and poetry have been reduced in our age to debated preferences for the right side or the left side of the brain. What the ancient and the contemporary arguments seem to forget, however, is that we have only one head in which poetic and scientific faculties must function. This is especially clear in our misunderstanding of what science is and how it works.

Despite a common assumption that science produces absolutes, it would be difficult to find a major scientist who would agree to any absolute claims in scientific discoveries. As Nobel prize biologist Sir Peter Medawar expresses it in his book *Pluto's Republic,* we should most properly think of science "as a logically articulated structure of justifiable *beliefs* about nature" (emphasis added). Since truth cannot ultimately be verified, scientists must actually get along with what is probable. The proof, insofar as it exists, comes from our *use* of a discovery, not from the absoluteness of the discovery expressed as a formula. As Socrates replied to a question about the truth of his depiction of the gods: "No, it is not true, but if you act as if it is true, you'll be doing the right thing." Rather than proof, science offers grounds for credibility and action.

Greater than our misunderstanding of the products and objectives of science, however, is our misunderstanding of the processes of science. Medawar explains that science begins with "an imaginative preconception of what might be true" and then critiques the preconception through experimentation. "It begins," he writes, "as a story about a Possible World—a story which we invent and criticize and modify as we go along, so that it ends up being, as nearly as we can make it, a story about real life." The first effort is to invent an imaginative picture that can be lived through or tested by experimentation. The purpose of this effort is to discover. The purpose of discovery is to create a closer harmony between the way we think about the world and the way the world works. Above all, the effort of discovery is heuristic. The process is both imaginative and without an absolute claim.

As we look forward to an age characterized by a quantum leap in generating and absorbing images, we lack confidence in the abilities of our imaginations. From where we stand right now, we will most likely let the new technology have its way. We will come to accept the conditions of a laissez-faire social contract, under which no one will have responsibility for what becomes of us. To counteract this tendency, we must begin to believe in our imaginations and understand how they work. Our survival kit must include the conviction that the imagination is a faculty for knowing things, for encountering and perceiving the real world, not a subjective fantasy world of our own making. The imagination is not just a faculty of fear, fantasy, unreality, and the dreams of the repressed. Rather, imagination is the faculty that stands between abstract thought and action. Imagination is where motivation and behavior meet. It is the imagination that will lift us out of the isolation and depression of the Confetti Generation and create a community. Consequently, we must know how it works and use it properly.

The Cognitive Imagination at Work

We generally accept that the way to understand and practice everything from basketball to cabinetmaking is to observe the skilled athletes and craftsmen at work and to imitate them. We function within our own limits. The speed, facility, strength, endurance, accuracy, and precision of the professionals astonish us, and in our own backyard or workroom we struggle along the paths they have trod before us. We must learn to operate in a similar way in the use and exercise of our imaginations.

We are all familiar with the "colorful imagination." The colorful imagination produces pictures or emotions without experience. It is the world of advertising, decoration, audiovisual aids, and nightmares. Coleridge called this talent "fancy," and it is generally what we mean by imagination. The colorful imagination, at its most honest and least exploitive, begins with a certain idea, meaning, or perception, which it elaborates on to persuade, excite, or entertain. If the product is successful, we can think our way back through the fabricated pictures to the original idea. Thus we have been taught to determine the "theme" of a work of art. What is to be learned or known can be reduced to a sentence. The rest is there as "a spoonful of sugar to help the medicine go down." Exercising this kind of imagination will simply fuel the Confetti Generation.

The works of art and science that grab our attention and force us to respond to them on their own terms, however, function quite differently. They are products of what we might call the "cognitive imagination." These products do not start off with meaning and then decorate it for our amusement. Rather, meaning develops in and through them. They themselves are instruments of knowledge and experience. Without them, we do not have access to the knowledge they embody. Whether such a product is *Hamlet* or "$E = mc^2$," it is by the end product and it alone that a certain meaning becomes clear. They are not reducible to simple statements of theme, and any attempt to do so immediately appears inadequate and incomplete. It is the process of producing these products of the imagination that we must identify and that we must exercise in our own limited way if we wish personally to survive the Confetti Generation.

"The essential feature of my creative thought," Albert Einstein once observed, "is the combinatory play of images." Thus the most recognizable scientific and abstract thinker of the twentieth century asserted that the central working mechanism of his thought was his imagination. He also provided us with the key to the process: "combinatory play." Since nightmares and advertising can also be described as a combinatory play of images, we must ask what makes them different from the process Einstein describes.

In the exercise of the cognitive imagination, the first step is the courageous assertion that there are such things as order, pattern, and meaning, both in the natural world and in the world of human behavior. What follows from this assertion is the urge to know and the energy to make sense. Both spring from faith in human existence. Einstein once reflected that his "God" was "order in the universe." If there is no order to be discovered, there is no sense in the effort.

Thus the first effort of the cognitive imagination begins with the belief that our thought can develop some kind of objective correlative for life both outside and inside ourselves, and that we can successfully construct such images. No thinker or creator claims or expects absolute success on this pilgrimage. In fact, failure of some sort is the norm. But like the judge who said that he knew pornography when he saw it, without being able to define it, so too can we comprehend things like love and justice, gravity and energy, the experience of adolescence, and the variations of defeat without defining them in abstract terms. We define them in combinations of images in stories and music, mathematics and physics. The goal of the search for comprehension is to achieve a greater harmony between the way we imagine the world of people, places, things, and events to be, and the way it is. Truth is the center around which the efforts of the cognitive imagination cohere—much as we perceive this to be the controlling principle of the pure sciences.

This goal of objective correlation is the opposite of subjectivism, charlatanism, kidding ourselves, or hiding behind carelessness—all so prevalent today in our rampant self-assertion, and themselves products of the colorful imagination. When we hold ourselves, our thoughts, and our lives to no standard other than opinion and convenience, we also demand nothing from the images we receive other than that they be stimulating. The cognitive imagination, on the other hand, demands objective communications and images and alters our expectations.

The second step in the exercise of the cognitive imagination is the combinatory play of images. It is a process. Our memory has already tested and stored images of experience that create the templates for new images or new combinations of old images. We work from the known to the as yet unknown by putting image next to image not only in the Sesame Street–like process of recognizing "same" and "different," but also in seeing if the images relate to each other and hang together. The method is metaphor and metamorphosis. It operates like a dance, in constant motion, one image distinct, then flowing into another to the point that we cannot tell the dancers from the dance.

The activity of the cognitive imagination is a constant process of identifying, isolating, sifting, weighing, and combining images, of consciously constructing clusters of images in which we become aware of experience, meaning, and the approximation of truth. We are all familiar with Santayana's dictum that those who cannot remember the past are condemned to repeat it. Only an active experiential memory of events and consequences in the past will avoid a repetition of history. We must grasp the pattern of current events, then sift and combine the multiple images of the past with those of the present to find the match that will inform and shape our choices for the future.

This same play of past and present images occurs when a mother sees her adult child behaving in a manner reminiscent of a time when the child was young. "Just like when you were two," she announces, usually to the adult child's chagrin. An image of behavior stored in the parent's memory has matched the image in current behavior, and meaning emerges, though the truth may hurt. This process of multiplying and extending images is what the cognitive imagination is all about.

The combinatory play of images is a participative, athletic, totally engaging process. There is nothing detached or abstract about it for it unites us with whatever becomes the object of our inquiry. Elizabeth Sewell describes this process in her book *The Orphic Voice* as "a way of using mind and body to build up dynamic structures (never fixed or abstract patterns) by which the human organism sets itself in relation to the universe and allows each side to interpret the other." In our example above, the mother connects two images of her child's behavior and actually feels the past and present life of the child in her current insight. At the same time she knows something about herself.

We are human organisms made up of mind and body, empowered with the gift of thought and the ability to act. Consequently, utilizing the cognitive imagination involves our total being at every step of the process. It means liberating ourselves from abstract thought and leaping over the scientific fallacy—what Coleridge called "the willing suspension of disbelief." By laying aside any inclination to abstraction, we come to understand behavior as behavior, and we recognize ourselves as participants in the process of change. Vitally engaging the cognitive imagination can rapidly produce new attitudes and behavior. We have always been integrated in mind and body, and we are seekers, not users. We are participants, not observers.

The third step in the exercise of the cognitive imagination is verification—how we decide that we've got it right. When do we temporarily stop the process and conclude that a particular image or combination of images is a useful approximation of the truth? At the turn of the century, French mathematician Henri Poincaré, whose formulas have now proven "practical" in the targeting of ICBMs, answered this question by saying that the right combinations are invariably identified by our response to their beauty. The most utilitarian combinations, he concluded, are precisely the most beautiful, and vice versa. Such is the truth of pure mathematics. In the theater it is, "There's Uncle Charlie!" When we poke the person next to us in delight while pointing to the stage, we are asserting and confirming the truth of a successful scene. This is the moment of recognition—the moment when the image or combination of images provides an experience that is itself the very recognition of experience.

These combinations of images are replicable and retrievable onstage, in a formula, or in our memories, and they stand the test of time. They hold up next to the facts of our own lives. Similar to confirmation in the empirical sciences, the cognitive imagination confirms fact—not fiction—by inviting experimental testing against life experiences. We prove fact by living just as we prove the theories of aerodynamics by flying, and flying more than once.

These disciplines of the cognitive imagination will be especially relevant to our survival during the Confetti Generation. No generation has ever gotten along without them, as is attested to by the seminal thinkers of past and present. But the precise relevance of the cognitive imagination for the Confetti Generation is in its individualized ability to control and derive meaning from the speed and inundation of images made possible by the new electronics. It gives us control at the control point. It is precisely designed to provide comprehension in the midst of the blizzard of data and images attacking our imaginations.

Furthermore, the cognitive imagination directly addresses the potential problems of remoteness and isolation, for it plugs us into the dynamic of life and the structures of the universe. It is participative. It is an exercise that includes us in a greater reality, rather than isolating us as the sole determinants of our singular reality. It is a technique for using our growing freedom to discover an approximation of truths that are more real than ourselves, and consequently can provide us with meaning and endurance.

Social Relevance

But is the cognitive imagination practical for daily living? We are not all geniuses like Einstein and Shakespeare. It seems too simple to suggest that the solution to the proliferation of data and images is to sort through them like the great creative minds of the past. Even if we accept that our epistemology has a direct correlation to our social character in a media-dominated society, philosophy is a long way away from our street-level perceptions and daily choices.

These are valid issues, and the place to begin may be with our philosophers, educators, and opinion makers, for the Confetti Generation will be their burden and generation as well as ours. Philosophy was not always remote from human life; our philosophers have made it so. From the Athenian marketplace to the origins of the university on the banks of the Seine, philosophers earned their living by attracting crowds and addressing their questions. Now they appear to address only their own questions in a closed society of their academic peers. To become more useful, philosophy as taught in our colleges must become more relevant to our living questions and become more widely studied.

What this leads to is social demands on our educational system. If philosophers should concentrate more on imaginative understanding of our problems rather than on techniques of linguistic analysis, more on ethics than on mathematics, so should all our educators. In common practice that adds up to the humanities, which are themselves products of the cognitive imagination. Teachers should invite entry to these creative works through the cognitive imagination. The opposite approaches of thematic abstraction, or exploring solely personal, opinionated responses, are unproductive. The cognitive imagination is athletic; exercising it is a craft. You train it by discipline, not by abstraction. You train it through practice and exercise, which is best done in and through the products of the cognitive imagination.

The cognitive imagination is also inherently dissatisfied with undigested experience. It craves a shape for information and data, even if that be nothing more than a chronological order for events, or the enumeration of possible reasons why an event took place, or a genuine story to explain fragmented news data. This is, in part, the job of the messengers, the reporters of events, who must present us with a way of thinking about information. It does not mean drawing a conclusion or voicing an opinion, any more than must drama or film. Too often, however, the news itself is nothing more than a series of disconnected events and a machine-gun chatter of data.

Undiluted data is also the plague of American business. We have lost our sense that those impersonal numbers actually represent human behavior, human productivity—human lives. The uneducated have a habit of feeling inadequate and insecure in the face of the educated. Data and numbers have become the new literacy, the new religious cult, the game of a new educated class. If so, this new class is a long way from the cognitive imagination, and it is not only in society's interest, but in their own interest, that they back off and reconsider what they are doing. The problem with being a gunfighter is that there is always somebody faster on the draw. In business terms, everybody reports to somebody; unless we find humanity in the numbers, we all lose our personhood.

In dealing with education, business, and the problems of daily living, we are no longer talking about philosophy and the humanities as academic disciplines. Rather, we are dealing with the cognitive imagination as a practical tool applicable to the important areas of family life. Consider diet, for example. We have learned a great deal about salt and cholesterol and preservatives, and this learning has changed our daily diets. We are also what we eat in communications terms, and our imaginative diet is a family matter. The cognitive imagination is no more directly concerned with censoring experiences out than it is with reliance on others for its entertainments. The cognitive imagination censors itself with boredom, and meaningless mayhem is boring.

If we celebrate the cognitive imagination, we will act it out in our family dialogue and behavior. Sex, violence, music videos, cartoons, computer graphics and numbers are not the problem, but *what we expect of them*. Our expectations of communications, data, and images determine family behavior as much as our expectations of food, exercise, table manners, arguments, and picking up our socks determine family behavior and personal character. We need not be geniuses to imitate how the great thinkers thought any more than we need to be professional athletes or woodworkers to approach basketball or cabinetmaking in a joyful and meaningful way.

From Self to Society

There is a difference between knowing the answer and knowing how and where to look. There is also a great difference between knowing how to look and consistently getting on with the work of discovery. The answer we have arrived at is a process. The cognitive imagination is a way of thinking and knowing; it is not itself coextensive with knowledge or meaningful behavior. It may represent something of a goal for the Confetti Generation, but it is not itself a goal. It is a description of how to proceed to find a way out of isolation and remoteness, a way to deal with the accelerated speed of images and the proliferation of data, and a way to find ourselves in an electronic world and make human connections. But the cognitive imagination is also a process and as such it lacks an ethic to keep it honest.

Originally, ethics, or *ethos,* identified the norms of behavior that defined or described a character or someone's consistent behavior. When we apply the words *ethics* or *character* today, we generally use them to describe governing principles that represent particularly acceptable norms of behavior. In most societies, whether religious or secular, it doesn't take long for such acceptable norms to be translated into demands and codified into law. The members of the society are made to perceive this ethical behavior as a means to achieve the higher goal of humanism, a just society, or eventual heavenly citizenship. In this context, we are perceived as the governed, and ethical behavior is the fulfillment of our obligations under the law. The situational ethics of the Autonomy Generation confound this traditional approach to ethical behavior.

The Confetti Generation will share in the struggle to arrive at mutually acceptable premises from which to construct norms of moral behavior. But our past history provides few clues for the construction of an ethic grounded in communications and relevant to communication. What we do have is distinctly impersonal, radically subject to equivocation, and generally inapplicable to the processes of knowing and communicating.

There is dialogue inherent in every thought, and particularly in the workings of the cognitive imagination. As we construct our ideas out of the combinatory play of images, we set up a framework of dialogue, of give-and-take, an I-Me or You-Me situation in our minds. Within an isolated mind we do not have any handy set of ethical principles for keeping this interior dialogue honest. More remarkable, however, we lack a general working ethic for exteriorizing this process in dialogue with others. We have laws and ethics relating to dishonesty and libel, but they are not very helpful in our daily communications where we seek meaning and relationships.

A useful approach to this problem that is particularly relevant for the Confetti Generation is contained in H. Richard Niebuhr's *The Responsible Self.* The concept of responsibility normally focuses on the self as agent or doer, and projects an image of us acting in the service of some goal or law. Niebuhr suggests, however, that "what is implicit in the idea of responsibility is the image of man-the-answerer," to complement older images of "man-the-maker" working toward some goal or "man-the-governed" meeting his obligations under the law. Man-the-answerer is "man engaged in dialogue, man acting in response to action upon him" for each of our actions is in some sense an answer to previous action upon us depending on how we interpreted that action. Understanding ourselves as responsive beings, Niebuhr says, is a useful intellectual stance that will bring to light "aspects of our self-defining conduct" so necessary for a generation intent upon defining themselves anyway.

It is not difficult to grasp the idea that all action is a response to other action, but what places the response in a moral arena is the effort of interpretation. The fact that an image enters our minds, whether from the natural world, memory, the subconscious, or a television tube, has no moral significance. Morality or ethics enters the equation the moment we interpret this image in preparation for our response. These moments of interpretation may be extended and complex, short and quick, or automatic and habitual. But we do interpret images, and that is the moment we enter into the combinatory play of images. Thus the first moral act and the foundation of an ethic rooted in communications is the effort of interpretation, the raising of the issues of meaning and significance.

As we are confrontd by the inundation of images and the increased imaginative encounters to be brought about by the new electronics, this insistence on interpretation will be the sine qua non of our ethical posture in exercising the cognitive imagination. Our question will not be "What is my end?" nor "What is the ultimate law?" Our questions during the Confetti Generation will be "What is going on?" and "What is being done to me?"

It requires effort to ask interpretive questions such as these as the ethical context for consciously committed responsive action. The basic effort is to continue the strenuous application of interpretation, to extend the combinatory play of images toward the goal of recognizable experience. The parallel effort is candor, the courage to be clear, the ability to admit insignificance as well as significance.

For Niebuhr, our actions can be considered responsible only insofar as they are reactions to interpreted actions upon us. Ethics relates to behavior. In our case that means that our ethics are centered not only on the serious and continuing effort of interpretation, but also on the conscious crafting of our responses to our interpretation of action upon us. When we respond, we are predicting a further response, and ideally our answer is subject to no other interpretation than that which we intend.

An ethic of responsible dialogue impels the effort of craftsmanship and preciseness in our shaping and juggling of images and our willingness to carefully and openly weigh the response that comes back. It applies to our interior as well as to our exterior dialogue and action. Whether an image enters our imagination from memory, from a television tube, or from another's behavior, we are ready to reply. We anticipate a continuation of the dialogue. We have sifted alternative images and metaphors to interpret what is going on or what is being done to us. Having made a choice, we design our response, the images we will create by our own words and deeds, with the intention of fitting into another's combinatory play of images. Our ethic will rest on the amount of care we use in designing this image so that as it stands alone outside ourselves and inside another, it can mean nothing other than what we intended. Our ethic will also depend on how open we are to the response we get, open to our failure to have been clear, and open to the adjustments and responses the next images invite.

An ethic of responsible communications is most easily explained in the context of one-on-one dialogue. But the same process applies to the dialogue of society. We are all part of the continuing discourse of people in our town and country, even the world. By participating in historical and evolutionary dialogue in a responsible way, we establish the foundation for social solidarity. Thus does the concept of the responsible self fit in with the concept of society communicating. We are not free of our culture. It is the context in which we meet and interpret people and events. It is this very inescapability of culture with its definitive character that gives our participation and contribution ethical weight. We cannot afford to break this fabric of society communicating. We are in dialogue, and how we act in response to the images inundating us will determine our character, both personal and social.

Our only hope for personal control and survival is to recognize the cognitive imagination as the place for research and discovery, to use its disciplines of inquiry and ethics, and to accept the logic behind its every product. We can confirm the objective relevance of each product by our response of recognition, by the admission that we have discovered something true about ourselves and our experiences that we did not know before and in a form that can be repeated and tested against growing experience. The products and processes of the thinking imagination are accessible and available to everyone—to anyone who can see or hear or feel or read or write. We are all, in fact, engaged in the process, whether we are aware of it or not. Becoming aware of this process and responsibly monitoring the ways the imagination deals with life in motion and translating that into behavior will be the primary challenge of the next generation.

The power of the imagination has led many people in history to fear it as far back as Plato, who would have banished poets from his ideal Republic. Our situation today is the direct opposite. We lack respect and show little concern for our imaginations. We have been able to shirk responsibility for our imaginations in part because of the collective imagination rooted in our culture, supported by the communications we all share. But in the next generation that collective imagination will fragment, and communications will become unique to each individual. At that point, we will be individually responsible for our imaginations. This is the kind of frightening lesson we all learn as we enter our first committed relationship. We suddenly discover that what we imagine the other person is saying by word and deed is critical and must be accurate. What we say and do in response is also critical. For the first time we cannot afford to be casual about what we imagine or how we respond to what we imagine. It is suddenly a serious business. This is the lesson we will—or will not—learn when the new electronics confronts us with the self-making fact of choice.

If we do not wish to experience the depression, isolation, and meaninglessness of the Confetti Generation, we must develop our imaginations as faculties of thought that understand that there are objective things that can be known only through the imagination and that these things must be known if we are to preserve our values and sense of self-worth. Buckminster Fuller has argued that the rampaging disorder of the world is balanced by man's creation of imaginative metaphysical order. This is not a job just for engineers, physicists, poets, and occasional philosophers. It is an individual task, requiring individual effort. Only at the individual level will we be able to energize the disciplines of imaginative order and ordering so as to deal successfully with the inundation, fragmentation, and isolation we are about to encounter in our communications.

Ethics and Regulation

In chapter 1 it was demonstrated that how we think about media determines both the communications products we receive and how we use them. What citizen consumers—individually and collectively—expect of communications is critical in the government and outcome of communications. Our expectations determine how we select, reject, and combine the images brought to us by technology as well as the communications we initiate ourselves as members of society.

In our experience with broadcast television, we missed two opportunities—the first by omission, the second by commission. In the first instance—a failure by omission—we failed to give any positive thrust to the "public interest" clauses that regulated broadcasting. With no articulate consensus on the positive aspects and expectations of television, the medium was effectively left free to follow its own purely economic and technological instincts. The second instance—a failure by commission—was the belated creation of public broadcasting, which, in fact, never intended to subscribe to the basic requirements of either the public arts or broadcasting technology. Consequently, public broadcasting failed to achieve the funding levels necessary for the creation and distribution of American cultural products. Its impact on the adult population has been minimal.

In light of these experiences there have been renewed and vigorous attempts to ensure that history does not repeat itself with regard to the emerging new electronic media. New moralists have arisen on both the right and the left who want the government, educators, churches, and families to use their various powers and influences consciously to shape the content and character of our communications.

Generally, however, Americans have always believed that their personal freedom, their democracy, and their economy are built on an ever-increasing abundance of communications and on the greatest possible number of competing communications choices simultaneously available to the individual citizen consumer. When it comes to communications, we are all Jeffersonian in our belief in the people's ability to decide for themselves—especially when that means *my* ability to decide for *myself*. Part of this freedom is the freedom to choose our own moralists from the right, center, or left. But it simultaneously precludes the imposition of our morality on others through government regulation of communications.

Beyond this deep-rooted cultural resistance, most of the

premises of government regulation are eroded by the very existence of the new electronic media. Foremost among these is that government regulation of electronic communications has always been premised on the scarcity of distribution channels. In broadcasting, it was the scarcity of spectrum space. In cable, it was the efficiency and convenience of limiting the use of public viaducts from telephone poles to underground tunnels. In either case the result was the public licensing of monopolies. However, cable destroyed the scarcity argument in broadcasting, and everything from satellite dishes to VCRs is eroding the principle as it applies to cable. Thus we are more likely to see a consensus developing around deregulation and a freer marketplace than around regulation.

On what regulatory subjects might a consensus be formed? One area is lying and deceit, which will translate into laws covering libel and fraud specifically related to the new electronics. Another is property rights, which will translate into new copyright statutes required by the new technology. But the key set of regulations, emerging out of the American love for individual freedom, will be new laws ensuring our rights to privacy. The more electronics intrudes into our lives, the more sacred the protection of personal information will appear. Finally, it is probable that a desire to ensure the benefits of abundance in communications will encourage regulations preventing the concentration of media ownership in too few hands.

The only possible conclusion for the future is that any regulations of communications will have little or nothing to do with the nature of the communications we do receive. We will be more than ever on our own. This is the same lesson we are learning from the impact of the new electronics on public broadcasting. The new electronics makes it absolutely clear that public broadcasting utilized the wrong technology for serving the high culture and elitist tastes it concentrated on with its programming. Now all of public broadcasting's products are available elsewhere and the free marketplace will determine precisely how interested we are in them. Ultimately we will determine our own communications diet, and we will not be able to rely on government regulation or support to ensure its nutritional content.

In the generation of new electronic media now emerging, control of communications will exist nowhere else but at its reception point. Social expectations, our social contract with communications and, consequently, our culture will depend on our individual actions more than at any time in history. Over the last thirty years we have followed a McLuhanesque rubric at this reception point. Burdened by the anxieties of other-direction, we have asserted that the interpretation of reality and even reality itself is purely subjective. Meaning is derived through free association of communicated images, and self-definition results from our spontaneous encounters. In a sense, we could afford to run the solipsistic risks of this strategy precisely because of the limits of our mass media and the homogeneity of our experiences.

During the Confetti Generation, however, we will be faced with the new burden of making sense out of experience not only *for ourselves* but *by ourselves*. The strategies we utilize at this point of reception and creation will determine our levels of isolation and meaninglessness, or community and significance. We will be defined by our epistemology. Our survival will depend on our choice of communications and on the methods we use for interpreting experience.

The Alternatives

The dynamics of the past few decades have produced some remarkable cultural shifts. The anxieties of an other-directed society drove many of us into a quest for self-actualization. In place of getting-and-spending in a homogeneous society, we emphasized the ecology of the psyche, with only a grudging acceptance of the outside world. We lived a molecular existence that treated the self as an object and provided no reference but the self. The tension soon became too great. Eventually we realized our need for a subjective relationship with at least one other human being if life were to be worth living. We moved toward a more integrated personalism, an ecology of the spirit rather than of the isolated psyche. Our point of view became less molecular. We reached for experience that included and needed others, but it remained a private contract, insisting on its own lyric voice and insights. Through this evolution, the Autonomy Generation shifted its focus from "I alone" to "I plus another," which is where we are today.

The question in doubt now is whether an Autonomy Generation with its previous emphasis on the psyche, or its newer emphasis on the spirit, possesses the cultural tools and is equipped with a personal survival kit adequate to the chaos that will inevitably arrive. It would appear that the only successful tools will revolve around a new kind of articulated humanism that we might call "incarnational ecology." It will be built on the recognition of the singular importance of the inner self and on the parallel recognition that we can only discover that self and articulate it in a finite world beyond ourselves. We might call it incarnational, because its fulcrum will be the cognitive imagination that can transform our personal worlds into the flesh of external reality. We might call it a new humanism, because it will be responsible both to our private selves and to humanity. Overall its emphasis will be on social creativity, a dramatic search for shared beliefs and actions that will overcome the fragmentation and isolation of communications. Our only practical hope in face of the eclectic solipsism of the Confetti Generation is to rediscover the thinking imagination and marry it to a dynamic, responsible, subjective approach to living in community. Only then will we change the equation, and we will do it only by changing ourselves.

Bibliography

Alfred P. Sloan Foundation. *On The Cable: The Television of Abundance*. Report of the Sloan Commission on Cable Communications. New York: McGraw-Hill, 1971.

Bell, Daniel. *The Coming of Post-Industrial Society: A Venture in Social Forecasting*. New York: Basic Books, 1973.

Carnegie Commission. *A Public Trust: The Report of the Carnegie Commission on the Future of Public Broadcasting*. New York: Bantam Books, 1979.

Crichton, Michael. *Electronic Life: How to Think About Computers*. New York: Knopf, 1983.

DeFleur, Melvin L. *Theories of Mass Communication*. New York: David McKay, 1966.

Galbraith, John Kenneth. *The Affluent Society*. Boston: Houghton Mifflin, 1976.

————. *The New Industrial State*. Boston: Houghton Mifflin, 1978.

Lasch, Christopher. *The Culture of Narcissism: American Life in an Age of Diminishing Expectations*. New York: Norton, 1978; Warner Books, 1979.

McLuhan, Marshall. *The Gutenberg Galaxy: The Making of Typographic Man*. Toronto: University of Toronto Press, 1962; New American Library, 1966.

————. *Understanding Media: The Extensions of Man*. New York: McGraw-Hill, 1965.

Maslow, Abraham H. *Motivation and Personality*. New York: Harper and Brothers, 1954.

————. *Toward a Psychology of Being*. New York: Van Nostrand, 1962.

May, Rollo. *Love and Will*. New York: Norton, 1969; Dell, 1974.

————. *The Meaning of Anxiety*. New York: Norton, 1950; Washington Square Press, 1979.

————, ed. *Existential Psychology*. New York: Random House, 1960.

Medawar, Peter. *Pluto's Republic*. New York: Oxford University Press, 1982.

Naisbitt, John. *Megatrends: Ten New Directions Transforming Our Lives*. New York: Warner Books, 1982.

Niebuhr, H. Richard. *The Responsible Self*. New York: Harper & Row, 1978.

Riesman, David, et al., *The Lonely Crowd*. New Haven: Yale, 1950.

Rogers, Carl R. *Client-Centered Therapy: Its Current Practice, Implications and Theory*. Boston: Houghton Mifflin, 1951.

————. *On Becoming a Person*. Boston: Houghton Mifflin, 1961.

Schramm, Wilbur. *Responsibility in Mass Communications*. New York: Harper, 1957.

————, ed. *Mass Communications*. Urbana: University of Illinois Press, 1960.

Sewell, Elizabeth. *The Orphic Voice: Poetry and Natural History*. New Haven: Yale, 1960.

Steinberg, Charles S., ed. *Mass Media and Communications*. New York: Hastings House, 1966.

Toffler, Alvin. *The Third Wave*. New York: Morrow, 1980; Bantam 1981.

Yankelovich, Daniel. *New Rules: Searching for Self-Fulfillment in a World Turned Upside Down*. New York: Random House, 1981; Bantam, 1982.

INDEX